OPTIMAL and ROBUST CONTROL

Advanced Topics with MATLAB®

OPTIMAL and ROBUST CONTROL

Advanced Topics with MATLAB®

Edited by
LUIGI FORTUNA
MATTIA FRASCA

CRC Press
Taylor & Francis Group
Boca Raton London New York

CRC Press is an imprint of the
Taylor & Francis Group, an **informa** business

CRC Press
Taylor & Francis Group
6000 Broken Sound Parkway NW, Suite 300
Boca Raton, FL 33487-2742

Printed in the United States of America on acid-free paper
Version Date: 20120104

International Standard Book Number: 978-1-4665-0191-1 (Hardback)

Library of Congress Cataloging-in-Publication Data

Fortuna, L. (Luigi), 1953-
Optimal and robust control : advanced topics with MATLAB / editors, Luigi Fortuna, Mattia Frasca.
p. cm.
Includes bibliographical references and index.
ISBN 978-1-4665-0191-1 (hardback)
1. Robust control--Mathematical models. 2. Linear systems. 3. Automatic control. I. Frasca, Mattia. II. Title.

TJ217.2.F67 2012
629.8--dc23 2011046677

Visit the Taylor & Francis Web site at
http://www.taylorandfrancis.com

and the CRC Press Web site at
http://www.crcpress.com

Dedicated to our wives

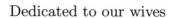

Contents

List of Figures

Preface

The scope of this book is to give to people with the knowledge of the main concepts of automatic control and signals and systems analysis a self-contained resource for advanced techniques both in linear system theory and in control design. This is achieved by including selected theoretical backgrounds, many numerical exercises and MATLAB® examples.

In this work we offer a complete and easy-to-read handbook of advanced topics in automatic control including the LQR and the H_∞ approach. Moreover our task has been to take advanced concepts of modeling analysis and present the essential items of the new technique of LMI in order to show how it can be considered as a unifying tool for system analysis and controller design. The robustness property of the closed control loop is the guideline of the book. The text deals with advanced automatic control techniques with particular attention to their *robustness*. Robustness means to guarantee the stability of a system in the presence of uncertainty. Uncertainty is due to the model itself or to the use of approximated models.

Many books on both the H_∞ control and the LQR control have been proposed since 1980. The LMI technique has become well known in the control community, and MATLAB® toolboxes to solve advanced control problems have been developed. However, often these subjects are presented only for specialists and such books are excellent sources for researchers and PhD students. This book, however, takes these topics and integrates them in an easy and concise way. This book is a compendium of many ordered subjects. For many proofs the reader is referred to the specific literature; however, many examples and many MATLAB® based solutions are included here to help the reader quickly acquire these skills.

The main task of the project has been to conceive a self-contained book of advanced techniques in automatic control that could be easily understood by those who have taken undergraduate courses in automatic control and systems. It is considered as the palimpsest of advanced modern topics in automatic control including an advanced set of analytical examples and MATLAB® exercises.

The book is organized into chapters structured as follows. The first chapter is an introduction to advanced control, the second cites some fundamental concepts on stability and provides the tools for studying uncertain systems. The third presents Kalman decomposition. The fourth chapter is on singular value decomposition of a matrix, given the importance of numerical techniques for systems analysis. The fifth, sixth and seventh chapters are on open-loop balanced realization and reduced order models. The eighth chapter presents the essential aspects of optimal control and the ninth discusses closed-loop balancing. The properties of passive systems and bounded real systems are the subjects of the tenth and eleventh chapters. Subsequently, the essential aspects of H_∞ control, the criteria for stability controls and LMI (Linear Matrix Inequalities) techniques commonly used in control systems design are dealt with. The book also includes numerous examples and exercises considered indispensable for coming to terms with the methodology and subject matter and a list of easy-to-find essential references.

This book is targeted at electrical, electronic, computer science, space and automation engineers interested in automatic control and advanced topics on automatic control. Mechanical engineers as well as engineers from other fields may also be interested in the topics of the book. The contents of the book can be learned autonomously by the reader in less than a semester.

Symbol List

Symbol Description

Symbol	Description
$\mathbb{R}$	the set of real numbers
$\mathbb{C}$	the set of complex numbers
$\mathbb{R}^n$	real vectors of n components
$\mathbb{C}^n$	complex vectors of n components
$\mathbb{R}^{m\times n}$	real matrices of dimensions $m \times n$
$\mathbb{C}^{m\times n}$	complex matrices of dimensions $m \times n$
$\Re e\ (x)$	real part a of a complex number $x = a + jb$
$\Im m\ (x)$	imaginary part b of a complex number $x = a + jb$
$\|x\|$	absolute value of a real number or the modulus of a complex number
$\\|x\\|$	norm of vector x
a_{ij}	coefficient of line i and column j of matrix A
sup	superior extreme of a set
inf	inferior extreme of a set
I	identity matrix of opportune dimensions
A^T	transpose of a matrix
A^*	conjugate transpose of $\mathrm{A} \in \mathbb{C}^{n\times n}$
$\mathrm{A} \otimes \mathrm{B}$	Kronecker product
$\mathrm{A} * \mathrm{B}$	Hadamard product (component by component product of two square matrices, i.e., $\mathrm{A} * \mathrm{B} = \{a_{ij}b_{ij}\}$)

Symbol	Description
$\det(\mathrm{A})$	determinant of matrix A
$\texttt{trace}(\mathrm{A})$	trace of matrix A
$\mathrm{P} > 0$	(semi-defined) defined positive matrix
$\mathrm{P} \geq 0$	semi-defined positive matrix
$\mathrm{P} < 0$	(semi-defined) defined negative matrix
$\mathrm{P} \leq 0$	semi-defined negative matrix
$\lambda_i(\mathrm{A})$	i-th eigenvalue of matrix A
$\sigma_i(\mathrm{A})$	i-th singular value of matrix A
$\rho(\mathrm{A})$	spectral radius (maximum eigenvalue) of matrix A
$S(\mathrm{A}, \mathrm{B}, \mathrm{C}, \mathrm{D})$	linear dynamical system $\dot{\mathbf{x}} = \mathrm{A}\mathbf{x} + \mathrm{B}\mathbf{u}$, $\mathbf{y} = \mathrm{C}\mathbf{x} + \mathrm{D}\mathbf{u}$
M_c	controllability matrix of a system
M_o	observability matrix of a system
σ_i	i-th singular value of a system
μ_i	i-th characteristic value of a system
$\texttt{RIC}(\mathrm{H})$	Riccati equation solution associated with Hamiltonian matrix H

SISO	single input single output system
MIMO	multi input multi output system

1

Modelling of uncertain systems and the robust control problem

CONTENTS

1.1 Uncertainty and robust control

In control system design the primary requirement is the asymptotic stability of controlled systems. In automatic control this is guaranteed through the design of an appropriate controller based on the nominal model of the process. However, in reality there may be several sources of uncertainty which make the nominal model inaccurate. *Robust control* deals explicitly with system uncertainties which are accounted for by differences between the real model and nominal model. Robust control guarantees controlled system performance (primarily asymptotic stability) when there are uncertainties.

The robust control issue can be summarized as: given a nominal process with acceptable interval values for perturbances, a controller should provide satisfactory performance in a closed-loop system for all the processes and "acceptable" perturbances.

As regards stability, the requisites of a robust control system should ensure:

1. closed-loop stability under nominal conditions;
2. closed-loop stability although there are uncertainties in the model.

As clarified below, uncertainties come in various forms, but, notwithstanding, an acceptable value range for possible uncertainties must be hypothesized within which interval control system performance is guaranteed. This can also explain how to establish if one controller is more robust than another: the greater the range of acceptable values for uncertainties, the more the controller can be considered robust.

The automatic control theory refers to the feedback control model represented in Figure 1.1 (unitary feedback) or in Figure 1.2 (non-unitary feedback). P represents the process and generally no assumptions are made about its lin-

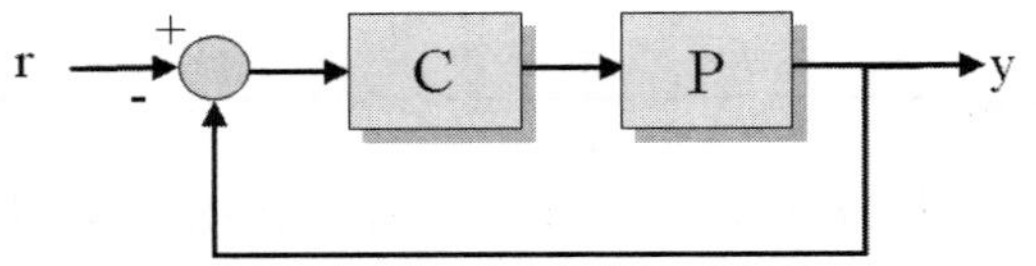

FIGURE 1.1
Unitary feedback control scheme.

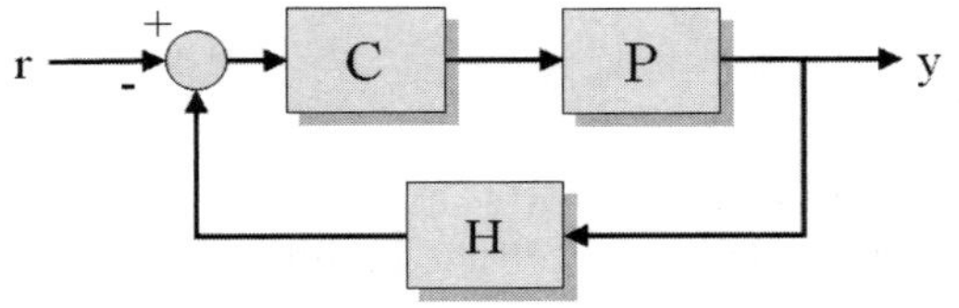

FIGURE 1.2
Feedback control scheme.

earity, assuming only it is a time-invariant system. C indicates the controller to be designed. The main issue is to design C so the closed-loop system is asymptotically stable. Furthermore, in addition to this basic specification, the control system can be required to respond to other criteria. In the case of robust control another important factor must be accounted for: the process model may have uncertainties.

For example, consider a linear process, described by transfer function $P(s) = \frac{1}{s^2(s+1)}$. Who can say for sure that the pole of this system is exactly -1? And, in the same way, on what basis could system gain be considered perfectly unitary? For this reason, in the theory of robust control, the system is indicated by $P(s) = \frac{g_1}{s^2(s+p_1)}$, inputting parameters which account for system uncertainties. $g_1 = 1$ and $p_1 = 1$ should be considered the parameters' nominal values, but they may have different values, whose variation range is usually known. This type of uncertainty concerns parameter values which characterize the system and so are defined *parametric uncertainty* or *structural uncertainty*. Examples of parametric uncertainty can be seen in daily life. Think of any size measurement (e.g., the length of a table): the measurement depends on the accuracy of the measuring instrument.

There are also other causes of parametric uncertainty. Parameter values may also vary depending on the operating conditions of the system. Think of a resistor heated by the Joule effect: electrical resistivity increases, modifying the value of that resistance. Or think of an airplane, as the fuel is consumed, the plane's total mass decreases.

Obviously, if parameter values change (even if in a predictable range), control system performance drops off compared to a nominal one (in fact, the design was built to nominal parameter values). The first goal of robust control is to ensure that, even with changing parameter values, the system's asymptotic stability is guaranteed.

Above, there was an example of parametric uncertainty about the coefficients of the transfer function of a linear time-invariant system. Now consider a nonlinear time-invariant system (with one input and output, for example), described by the equations:

$$\begin{aligned} \dot{\mathbf{x}} &= \mathrm{f}(\mathbf{x}) + \mathrm{g}(\mathbf{x})u \\ y &= \mathrm{h}(\mathbf{x})u \end{aligned} \tag{1.1}$$

with $\mathbf{x} \in \mathbb{R}^n$ (state variables), $u \in \mathbb{R}$ (system input) and $y \in \mathbb{R}$ (system output), $\mathrm{g} : \mathbb{R}^n \to \mathbb{R}^n$, $\mathrm{f} : \mathbb{R}^n \to \mathbb{R}^n$ and $\mathrm{h} : \mathbb{R}^n \to \mathbb{R}^n$. With robust control, parametric uncertainty in the model is highlighted by using the parameters α, β and γ:

$$\begin{aligned} \dot{\mathbf{x}} &= \mathrm{f}(\mathbf{x}, \alpha) + \mathrm{g}(\mathbf{x}, \beta)u \\ y &= \mathrm{h}(\mathbf{x}, \gamma)u \end{aligned} \tag{1.2}$$

In this case, $\bar{\alpha}$, $\bar{\beta}$ and $\bar{\gamma}$ indicate the nominal values of the parameters α, β and γ. Robust control for system (1.2) means designing C to guarantee closed-loop asymptotic stability, given the variation spans of parameters α, β and γ (generally, vectors of arbitrary size).

Similarly, one can consider linear systems in state-space form or other nonlinear models. Especially, for a linear time-invariant system in state-space form:

$$\begin{aligned} \dot{\mathbf{x}} &= \mathrm{A}\mathbf{x} + \mathrm{B}\mathbf{u} \\ \mathbf{y} &= \mathrm{C}\mathbf{x} + \mathrm{D}\mathbf{u} \end{aligned} \tag{1.3}$$

with $\mathbf{x} \in \mathbb{R}^n$, $\mathbf{u} \in \mathbb{R}^m$, $\mathbf{y} \in \mathbb{R}^p$, $\mathrm{A} \in \mathbb{R}^{n\times n}$, $\mathrm{B} \in \mathbb{R}^{n\times m}$, $\mathrm{C} \in \mathbb{R}^{p\times n}$, $\mathrm{D} \in \mathbb{R}^{p\times m}$, coefficients of A, B, C and D vary in certain intervals. What is required of robust control, once the controller is designed for nominal parameter values, is that asymptotic stability is guaranteed even when the parameters are not nominal.

The linear time-invariant systems can also be described on a more compact form with the *realization matrix*:

$$\mathrm{R} = \begin{bmatrix} \mathrm{A} & \mathrm{B} \\ \mathrm{C} & \mathrm{D} \end{bmatrix} \tag{1.4}$$

If the number of inputs equals the number of system outputs ($p = m$) then matrix R is square ($\mathrm{R} \in \mathbb{R}^{(n+m)\times(n+m)}$). The eigenvalues of this matrix can be easily shown not to depend on the reference system. In fact a state transformation leads to:

$$\tilde{\mathrm{R}} = \mathrm{T}^{-1}\mathrm{R}\mathrm{T} \tag{1.5}$$

which is clearly a similitude relation.

To verify if a system is minimal is to calculate the eigenvalues of matrix R which are represented by $\bar{\lambda}_1, \bar{\lambda}_2, \ldots, \bar{\lambda}_{n+m}$, whereas the eigenvalues of A by $\lambda_1, \lambda_2, \ldots, \lambda_n$. The system is minimal if none of eigenvalues of A coincide with those of R.

There is another type of uncertainty which is more difficult to deal with than parametric uncertainties. Called *structural uncertainty* or *unstructured uncertainty*, it concerns model structure. Consider again the example of the process described by $P = \frac{1}{s^2(s+1)}$, structural uncertainty takes into account the possibility that the modelling did not account perhaps for an additional pole and an additional zero in process transfer function:

$$P = \frac{(\frac{s}{\alpha_0} + 1)}{s^2(s+1)(\frac{s}{\alpha} + 1)}$$

So, structural uncertainty derives from incorrect modelling, which overlooked perhaps a dynamic. Neglecting a modelling dynamic is actually very common and can have consequences for closed-loop stability.

Example 1.1

Consider system $P(s) = \frac{1}{(s+1)^2}$. The transfer function of the closed-loop system (Fig. 1.1) is given by $M(s) = \frac{Y(s)}{R(s)} = \frac{C(s)P(s)}{1+C(s)P(s)}$. Using a simple proportional controller $C(s) = k$, you obtain $M(s) = \frac{k}{(s+1)^2+k}$ and the closed-loop system is asymptotically stable $\forall k > 0$.

For example, if $C(s) = 100$, notice that this controller is robust to parametric variations of position of the pole of $P(s)$ with double multiplicity. In fact, if instead of $p = -1$, it was $p = -\alpha$, the controller $C(s) = 100$ would continue to guarantee closed-loop asymptotic stability to a large value of the parameter ($\alpha > 0$). In the case of structural uncertainty, the scenario is different. Suppose that there is an uncertainty perhaps on system order, so that $P(s) = \frac{1}{(s+1)^3}$. The characteristic closed-loop equation is given by $(s+1)^3 + k = 0$, i.e., $s^3 + 3s^2 + 3s + 1 + k = 0$. Applying the Routh criterion we note that in this case the system is asymptotically stable if $0 < k < 8$.

Controller $C(s) = 100$, designed for the nominal system $P(s) = \frac{1}{(s+1)^2}$, no longer guarantees asymptotic stability because of the structural uncertainty of an additional pole at -1. The conclusion is that the consequences of uncertainty on system order, having overlooked a dynamic (which could also be a dynamic parasite which triggers in certain conditions), can be very important.

Defining parametric uncertainty is done by expressing a parameter variation range. For example, given a linear time-invariant system described by its transfer function

$$P(s) = \frac{N(s)}{D(s)} = \frac{b_0 s^m + b_1 s^{m-1} + \ldots + b_{m-1}s + b_m}{s^n + a_1 s^{n-1} + \ldots + a_{n-1}s + a_n}$$

once the minimum and maximum values of the various coefficients are assigned, parametric uncertainty is completely characterized:

$$\begin{array}{c} a_1^m \leq a_1 \leq a_1^M \\ \cdots \\ a_n^m \leq a_n \leq a_n^M \\ b_0^m \leq b_0 \leq b_0^M \\ \cdots \\ b_m^m \leq b_m \leq b_m^M \end{array} \tag{1.6}$$

If there is structural uncertainty we assume $P(s) = P_0(s) + \Delta P(s)$, where $P_0(s)$ represents the nominal model and $\Delta P(s)$ represents uncertainty. This structural uncertainty is *additive*. Structural uncertainty can also be *multiplicative* $P(s) = P_0(s)\Delta P(s)$. Structural uncertainty is measured by a norm of the transfer function norm or the transfer matrix. We will see how to define such a norm in Chapter 10.

The greater the required robustness, the greater the precision required to model the process. A particularly important role in robust control is testing. The popularity of personal computers now makes it possible to perform tests through numerical simulations even when the robust control problem has no analytical solutions. Numerical simulations help verify operation and controller performance in the variation interval indicated by the uncertainty.

The most critical issue is structural uncertainty. Suppose we consider a horizontal beam fixed at one end and subject to a stimulus at the other. This system has distributed parameters and could be modelled by an infinite series of mass-spring systems. The free-end deflection has infinite modes which can be described with a transfer function of type $G(s) = \sum_{i=0}^{\infty} G_i(s)$, where $G_i(s)$ is the transfer function describing the mode associated with the n-th section. However, this approach is not so easy to carry out and so an approximation $G(s) = G_0(s) + \Delta G(s)$ is used which takes into account the most important modes (modelled by $G_0(s)$). Approximation is necessary for the design of a non-distributed controller (note too that additive uncertainty helps model the high frequency dynamic neglected by the approximation). In these cases only the dominant part is considered, but it is essential that the controller is robust enough for any uncertainty in the dynamic not covered by the model. This introduces another very important issue treated below: how to figure out reduced order models which can allow to deal in a more simply way with the control problem. It seems clear that this operation introduces an error which can be evaluated and taken into account in the robust controller design.

Now, let's consider the case of the linear time-invariant systems described by equations (1.3). Under opportune hypotheses, a controller can be designed using a control law which operates on state variables (the linear state regulator $\mathbf{u} = -\mathrm{k}\mathbf{x}$). Remember, should not all the state variables be accessible, they would have to be re-constructed through an asymptotic state observer. This common technique usually includes the design of a linear regulator and an asymptotic observer which together constitute the compensator.

When designing a linear state regulator, the system must be completely controllable (i.e., matrix $\mathrm{M}_c = \begin{bmatrix} \mathrm{B} & \mathrm{AB} & \mathrm{A}^2\mathrm{B} & \dots & \mathrm{A}^{n-1}\mathrm{B} \end{bmatrix}$ has maximum rank), whereas designing an asymptotic observer needs the system to be completely observable (i.e., matrix $M_o = \begin{bmatrix} C & CA & CA^2 & \dots & CA^{n-1} \end{bmatrix}^T$ has maximum rank).

Furthermore, note that, since the observer is a dynamical system (linear and time-invariant) of order n, then the compensator obtained is a dynamical system of order n. High order systems require the design of high order compensators. So, now, the problem of formulating a lower order model also includes the design of a compensator. To do this, there are two different techniques. The first is to design a compensator and then build the lower order model. The second is to design the compensator directly on the lower order model (e.g., considering $P_0(s)$ and neglecting $\Delta P(s)$).

It should not be forgotten that any uncertainties regard process P and not controller C. It is supposed that it will be the designer who builds the controller, so any uncertainties about its parameters are minimal. More recently, it has been discovered that compensators designed with the theory of robust control are fragile, that is, that compared to the compensator parameters, robustness is poor, so even minimal uncertainties about compensator coefficients can lead to system de-stabilization. At the heart of the compensator issue is that when it's designed to be robust against the uncertainties in the model of process P, it remains fragile regarding the uncertainties in the model of controller C.

This chapter has briefly described the main robust control issues to be dealt with in subsequent chapters. Before that, certain subjects will be briefly presented which are the pillars of systems theory (e.g., stability). This approach was decided on for two reasons: first, is the importance of these subjects to robust control, and, second, is that undergraduate courses are organized on two levels. The course which this book is part of is a Master's course for students who have likely dealt with automation previously. So, this makes a brief discourse on systems theory necessary.

1.2 The essential chronology of major findings into robust control

Parameterization of stabilizing controllers (Youla, 1976).
Poor robustness of LQG controllers (Doyle, 1978).
Formulating H_∞ problems for SISO systems (Zames, 1981).
Balanced realizations (Moore, 1981).
Definition of μ (Doyle, 1982).
Definition of multivariable stability margins (Safonov, 1982).
H_∞ synthesis algorithms for systems in state-space form: high order control (Doyle, 1984).
First book on robust control ((Francis, 1987).

H_∞ synthesis algorithms for systems in state-space form: low order control (Doyle, Glover, Khargonekar, Francis, 1988).

2

Fundamentals of stability

CONTENTS

This chapter deals briefly with the problem of stability, the main requirement of a control system. In going over some basics in systems analysis, the emphasis is on some advanced mathematical tools (eg. Lyapunov equations) which will be useful below.

2.1 Lyapunov criteria

The equilibrium points of an autonomous dynamical system $\dot{\mathbf{x}} = f(\mathbf{x})$ with $\mathbf{x} = \begin{bmatrix} x_1 & x_2 & \dots & x_n \end{bmatrix}^T$ and $f(\mathbf{x}) = \begin{bmatrix} f_1(\mathbf{x}) & f_2(\mathbf{x}) & \dots & f_n(\mathbf{x}) \end{bmatrix}^T$ can be calculated by $\dot{\mathbf{x}} = 0$ (i.e., solving the systems of equations $f(\mathbf{x}) = 0$). Generally, the equations are nonlinear and may have one or more solutions, whereas for autonomous linear systems, the equilibrium points can be calculated by $A\mathbf{x} = 0$, which has only one solution ($\mathbf{x} = 0$) if $\det(A) \neq 0$.

So, nonlinear systems may have more than one equilibrium point. In addition, each of these equilibrium points has its own stability characteristics. For example, think of a pendulum. It has two equilibrium points as shown in Figure 2.1. Only the second of the two equilibrium points is stable. If the pendulum were to start from a position near equilibrium point (a), it would not return to its equilibrium position, unlike what happens at point (b). For this reason in nonlinear systems, stability is a property of equilibrium points, and not of the whole system. The stability of an equilibrium point can be studied through the criteria introduced by the Russian mathematician Aleksandr Mikhailovich Lyapunov (1857-1918).

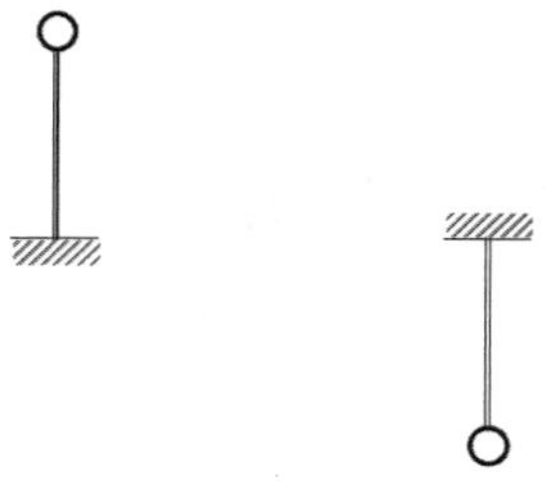

FIGURE 2.1
Equilibrium points of a pendulum.

The first of Lyapunov's criteria defines stability of an equilibrium point. It says that if a system starts from initial conditions very close to a stable equilibrium point, the state evolution is confined to the neighborhood of that equilibrium point. If the initial conditions $\mathbf{x}(0) = \mathbf{x}_0$ are close to equilibrium, the response obtained by $\mathbf{x}_0$, i.e., *perturbed response*, is sufficiently close to equilibrium point $\bar{\mathbf{x}}$ (the *nominal response*, i.e., the motion obtained starting exactly from $\bar{\mathbf{x}}$, is in fact constant over time and equal to the equilibrium point). To measure the distance between perturbed and nominal motion in finite space, any vector norm can be used. The criterion is formally expressed by:

Definition 1 (Lyapunov I criterion) *Equilibrium point $\bar{\mathbf{x}}$ of dynamical system $\dot{\mathbf{x}} = f(\mathbf{x})$ ($f(\bar{\mathbf{x}}) = 0$) is defined as stable when:*

$$\forall \varepsilon > 0,\ \exists \delta > 0 \textit{ such that } \forall \mathbf{x}_0 \textit{ with } \parallel \mathbf{x}_0 - \bar{\mathbf{x}} \parallel < \delta \textit{ one has}$$
$$\parallel \mathbf{x}(t) - \bar{\mathbf{x}} \parallel < \varepsilon \textit{ for } \forall t > 0$$

Stability is said to be asymptotic, if this additional condition applies:

$$\lim_{t \to +\infty} \mathbf{x}(t) = \bar{\mathbf{x}} \tag{2.1}$$

In this case, the perturbed motion not only stays in the neighborhood of the equilibrium point, but asymptotically tends to it.

Definition 2 (Asymptotically stable equilibrium point) *Equilibrium point $\bar{\mathbf{x}}$ of dynamical system $\dot{\mathbf{x}} = f(\mathbf{x})$ ($f(\bar{\mathbf{x}}) = 0$) is defined as asymptotically stable if it is stable and furthermore if condition (2.1) holds.*

Before discussing Lyapunov's second criterion, some preliminary notions need introducing.

Note first that, without loss of generality, it is possible to consider $\bar{\mathbf{x}} = 0$, since it is always possible to shift the generic equilibrium point $\bar{\mathbf{x}}$ back to 0 by variable translation.

Now, the concept of positive definite functions needs to be introduced. Let's consider a neighborhood $\Omega \subseteq \mathbb{R}^n$ of point $\bar{\mathbf{x}} = 0$.

Definition 3 (Positive definite function) *Function $V(\mathbf{x}) : \mathbb{R}^n \to \mathbb{R}$ is positive definite in $\bar{\mathbf{x}}$ if the following hold:*

1. $V(0) = 0$
2. $V(\mathbf{x}) > 0 \ \forall \mathbf{x} \in \Omega, \ \mathbf{x} \neq 0$

A function is positive semi-definite if $V(0) = 0$ and if $V(\mathbf{x}) \geq 0 \ \forall \mathbf{x} \in \Omega$. A function is negative definite if $V(0) = 0$ and if $V(\mathbf{x}) < 0 \ \forall \mathbf{x} \in \Omega, \ \mathbf{x} \neq 0$. Finally, a function is negative semi-definite if $V(0) = 0$ and if $V(\mathbf{x}) \leq 0$ $\forall \mathbf{x} \in \Omega$.

Theorem 1 (Lyapunov II criterion) *The equilibrium $\bar{\mathbf{x}} = 0$ of dynamical system $\dot{\mathbf{x}} = f(\mathbf{x})$ is stable if there is a scalar function of the state, the so-called Lyapunov function, which has the following properties:*

1. *$V(\mathbf{x})$ is positive definite for $\bar{\mathbf{x}} = 0$*
2. *$\dot{V}(\mathbf{x}) = \frac{dV(\mathbf{x})}{dt}$ is negative semi-definite for $\bar{\mathbf{x}} = 0$*

The stability is asymptotic if $\dot{V}(\mathbf{x})$ is negative definite.

From a physical point of view, the Lyapunov function represents system energy. If the energy grows over time, it means that the state variables grow (thus diverge). Conversely, if the energy tends to zero, the system is dissipating energy.

Note that many Lyapunov functions satisfy the first condition of Lyapunov II criterion; the difficulty lies in verifying the second condition.

2.2 Positive definite matrices

For linear systems, a Lyapunov function allowing to assess system stability is easily found. In fact, for nonlinear systems, there is no general procedure for finding Lyapunov functions, while for linear systems there is. Effectively, Lyapunov second criterion for linear systems can be simplified. Before introducing the criterion, firstly, positive definite matrices are defined.

Given matrix $\mathrm{P} \in \mathbb{R}^{n \times n}$, a quadratic form can be associated to it: $V(\mathbf{x}) = \mathbf{x}^T \mathrm{P} \mathbf{x}$, which is clearly a scalar function of vector $\mathbf{x} \in \mathbb{R}^n$.

Definition 4 *Matrix P is positive definite if the quadratic form associated to it is positive definite.*

Likewise, it is possible to define negative definite, positive semi-definite and negative semi-definite matrices.

Now, we discuss how to determine if a matrix is positive definite. It turns out that this depends only on the so-called *symmetrical part* of the matrix. In fact, matrix P can always be written as the sum of two matrices, one symmetrical and one asymmetrical, named *symmetrical part* P_s and *antisymmetrical part* P_{as} of matrix:

$$\mathrm{P} = \mathrm{P}_s + \mathrm{P}_{as}$$

where the symmetrical part is

$$\mathrm{P}_s = \frac{1}{2}(\mathrm{P} + \mathrm{P}^T)$$

and the antisymmetrical part is

$$\mathrm{P}_{as} = \frac{1}{2}(\mathrm{P} - \mathrm{P}^T).$$

Clearly, only the symmetrical part of a matrix determines if a matrix is positive definite:

$$V(\mathbf{x}) = \mathbf{x}^T \mathrm{P} \mathbf{x} = \mathbf{x}^T \mathrm{P}_s \mathbf{x}$$

The asymmetrical part contributes nothing:

$$V(\mathbf{x}) = \mathbf{x}^T \mathrm{P} \mathbf{x} = p_{11}x_1^2 + \ldots + p_{nn}x_n^2 + (p_{12} + p_{21})x_1x_2 + (p_{13} + p_{31})x_1x_3 + \ldots$$
$$\ldots + (p_{n-1,n} + p_{n,n-1})x_{n-1}x_n$$

For the other part of the definition of P_s:

$$\mathrm{P}_s = \begin{bmatrix} p_{11} & \frac{p_{21}+p_{12}}{2} & \ldots & \frac{p_{n1}+p_{1n}}{2} \\ \frac{p_{21}+p_{12}}{2} & p_{22} & \ldots & \frac{p_{n2}+p_{2n}}{2} \\ \vdots & \vdots & & \vdots \\ \frac{p_{n1}+p_{1n}}{2} & \frac{p_{n2}+p_{2n}}{2} & \ldots & p_{nn} \end{bmatrix}$$

from which it follows that $V(\mathbf{x}) = \mathbf{x}^T \mathrm{P}_s \mathbf{x}$.

Let's recall some properties of symmetrical matrices that will be useful below.

- The eigenvalues of a symmetrical matrix are real.
- A symmetrical matrix is always diagonalizable. This means that there is always a set of linear independent and orthogonal eigenvectors, such that $V_j^T V_i = 0$ if $i \neq j$. For symmetrical matrices it is always possible to find a set of orthonormal eigenvectors such that if

$\mathrm{V} = \begin{bmatrix} V_1 & V_2 & \dots & V_n \end{bmatrix}$ then:

$$\mathrm{V}^T\mathrm{V} = \mathrm{I}$$

$$\mathrm{V}^T = \mathrm{V}^{-1}$$

and

$$P = \mathrm{V}\Lambda\mathrm{V}^T$$

From this, we obtain a significant finding for positive definite matrices. Consider the quadratic form $V(\mathbf{x})$ of P, which equals:

$$V(\mathbf{x}) = \mathbf{x}^T\mathrm{P}_s\mathbf{x} = \mathbf{x}^T\mathrm{V}\Lambda\mathrm{V}^T\mathbf{x}$$

Consider the state transformation $\tilde{\mathbf{x}} = \mathrm{V}^T\mathbf{x}$, then:

$$V(\tilde{\mathbf{x}}) = \tilde{\mathbf{x}}^T\Lambda\tilde{\mathbf{x}} \tag{2.2}$$

Clearly, $V(\mathbf{x})$ and $V(\tilde{\mathbf{x}})$ have the same properties, but understanding if $V(\tilde{\mathbf{x}})$ is positive definite is much simpler. In fact, if all the eigenvalues of P_s are positive, then $V(\tilde{\mathbf{x}})$ is positive definite. This result can be summarized in the theorem below.

Theorem 2 *A symmetrical matrix* P *is positive definite if all its eigenvalues are positive.*

The matrix is positive semi-definite if all its eigenvalues are non-negative and there is at least one null eigenvalue. Moreover, P is negative definite if all its eigenvalues are negative and negative semi-definite if all its eigenvalues are non-positive and at least one of them is null. If the matrix is not symmetrical, the same conditions apply except that the eigenvalues must be from the symmetrical part of the matrix.

The following theorem is useful for determining if a symmetrical matrix is positive or not without calculating the eigenvalues.

Theorem 3 (Sylvester test) *A symmetrical matrix* P *is positive definite if and only if all of its n principal minors* $\mathrm{D}_1, \mathrm{D}_2, \dots, \mathrm{D}_n$ *are positive, i.e.:*

$$\mathrm{D}_1 = p_{11} > 0; \mathrm{D}_2 = \det\begin{bmatrix} p_{11} & p_{12} \\ p_{21} & p_{22} \end{bmatrix} > 0; \dots; \mathrm{D}_n = \det\mathrm{P} > 0 \tag{2.3}$$

MATLAB® exercise 2.1

Consider $P = \begin{bmatrix} 1 & 2 & 5 \\ 2 & 5 & -1 \\ 5 & -1 & 0 \end{bmatrix}$. Since the matrix is symmetrical, the Sylvester test can be applied to determine if it is positive definite or not.

Let us first define matrix P in MATLAB with the command:

```
>> P=[1 2 5; 2 5 -1; 5 -1 0]
```

and then compute D_1, D_2 and D_3 as follows:

```
>> D1=det(P(1,1))
>> D2=det(P(1:2,1:2))
>> D3=det(P)
```

One obtains: $D_1 = 1$, $D_2 = 1$ and $D_3 = -146$. Since $D_3 < 0$, P is not positive definite.

2.3 Lyapunov theory for linear time-invariant systems

The Lyapunov second criterion for linear time-invariant systems concerns the *system stability.* In fact, in linear systems, since all the equilibrium points have the same stability properties, it is possible to refer to the stability of the system. The criterion is expressed by the following theorem.

Theorem 4 (Lyapunov II criterion for linear time-invariant systems) *A linear time-invariant system* $\dot{\mathbf{x}} = A\mathbf{x}$ *is asymptotically stable if and only if for any positive definite matrix* Q *there exists a unique positive definite matrix* P*, which satisfies the following equation (the so-called Lyapunov equation):*

$$A^T P + PA = -Q \tag{2.4}$$

Equation (2.4) is a *linear* matrix equation, because the unknown P appears with a maximum degree equal to 1.

For simplicity, usually symmetrical Q is chosen. Here too P is symmetrical:

$$(A^T P + PA)^T = -Q^T \Rightarrow P^T A + A^T P^T = -Q^T$$
$$\Rightarrow P^T A + A^T P^T = -Q \Rightarrow P^T = P$$

One way of solving Lyapunov's equation is by *vectorization.*

Example 2.1

Consider an example with n equal to 2:

$$A = \begin{bmatrix} -2 & 0 \\ 0 & -5 \end{bmatrix}$$

Since the Lyapunov criterion states that equation (2.4) should hold for any positive definite Q, then let's choose Q = I (indeed, it suffices to prove it for one Q matrix, to show that it holds for any).

Consider matrix P made up of three unknowns p_{11}, p_{12} and p_{22}:

$$P = \begin{bmatrix} p_{11} & p_{12} \\ p_{12} & p_{22} \end{bmatrix}.$$

Solving the Lyapunov equation means solving a system of three equations with three unknowns; by substituting in equation (2.4):

$$\begin{bmatrix} -2 & 0 \\ 0 & -5 \end{bmatrix} \begin{bmatrix} p_{11} & p_{12} \\ p_{12} & p_{22} \end{bmatrix} + \begin{bmatrix} p_{11} & p_{12} \\ p_{12} & p_{22} \end{bmatrix} \begin{bmatrix} -2 & 0 \\ 0 & -5 \end{bmatrix} = \begin{bmatrix} -1 & 0 \\ 0 & -1 \end{bmatrix}$$

$$\Rightarrow \begin{cases} 4p_{11} = 1 \\ 7p_{12} = 0 \\ 10p_{22} = 1 \end{cases}$$

The three equations with three unknowns can be re-written in matrix form:

$$\begin{bmatrix} 4 & 0 & 0 \\ 0 & 7 & 0 \\ 0 & 0 & 10 \end{bmatrix} \begin{bmatrix} p_{11} \\ p_{12} \\ p_{22} \end{bmatrix} = \begin{bmatrix} 1 \\ 0 \\ 1 \end{bmatrix}$$

This leads to: $P = \begin{bmatrix} \frac{1}{4} & 0 \\ 0 & \frac{1}{10} \end{bmatrix}$. Matrix P is positive definite, so the system is asymptotically stable.

Notice that, since in example 2.1 matrix A is diagonal, concluding that the system is stable is immediate from the inspection of the eigenvalues of A which are all in the closed left-hand half of complex plane. Generally, testing system stability via the Lyapunov second criterion is not numerically very efficient, since the first Lyapunov equation has to be solved and then the positive definiteness of the solution has to be verified. Tests on the eigenvalues of A or criteria such as the Routh criterion are more direct and efficient. If A is known, the Lyapunov second criterion is not the most efficient method to test the stability of the linear system. However, it turns out that the Lyapunov second criterion is a powerful theoretical tool for proving system stability.

Generally, if Q is not symmetrical the vectorization is:

$$M \begin{bmatrix} p_{11} \\ p_{12} \\ p_{13} \\ \vdots \\ p_{nn} \end{bmatrix} = - \begin{bmatrix} q_{11} \\ q_{12} \\ q_{13} \\ \vdots \\ q_{nn} \end{bmatrix}$$

where $M \in \mathbb{R}^{n^2 \times n^2}$.

Proof 1 *(Proof of sufficiency of the second criterion of Lyapunov for linear time-invariant systems). This proves the sufficient part of the Lyapunov criterion for linear time-invariant systems, so if* $\forall$ Q *positive definite, there exists a matrix* P *satisfying Lyapunov equation (2.4), then the system is asymptotically stable.*

Consider the Lyapunov quadratic function $V(\mathbf{x}) = \mathbf{x}^T P \mathbf{x}$. *Since* P *is a positive definite matrix,* $V(\mathbf{x})$ *is a positive definite function. The derivate against time of this function is:*

$$\dot{V}(\mathbf{x}) = \dot{\mathbf{x}}^T \mathrm{P}\mathbf{x} + \mathbf{x}^T \mathrm{P}\dot{\mathbf{x}}$$

Since $\dot{\mathbf{x}} = \mathrm{A}\mathbf{x}$, *then:*

$$\dot{V}(\mathbf{x}) = \mathbf{x}^T \mathrm{A}^T \mathrm{P}\mathbf{x} + \mathbf{x}^T \mathrm{P}\mathrm{A}\mathbf{x} =$$

$$= \mathbf{x}^T (\mathrm{A}^T \mathrm{P} + \mathrm{P}\mathrm{A})\mathbf{x} = -\mathbf{x}^T \mathrm{Q}\mathbf{x}$$

Since Q *is a positive definite matrix,* $\dot{V}(\mathbf{x})$ *is a negative definite function. Function* $V(\mathbf{x}) = \mathbf{x}^T \mathrm{P}\mathbf{x}$ *satisfies the hypotheses of Lyapunov II criterion for dynamical systems and thus the equilibrium point (and so the system) is asymptotically stable.*

Before proving the necessary part of the theorem, an important property of positive definite matrices should be highlighted.

Theorem 5 *If* $\mathrm{Q} \in \mathbb{R}^{n \times n}$ *is a positive definite matrix and* $\mathrm{M} \in \mathbb{R}^{n \times n}$ *is a full rank matrix, then also* $\mathrm{M}^T\mathrm{Q}\mathrm{M}$ *is a positive definite matrix.*

Proof 2 *The quadratic form associated to* $\mathrm{M}^T\mathrm{Q}\mathrm{M}$ *is*

$$V(\mathbf{x}) = \mathbf{x}^T \mathrm{M}^T \mathrm{Q}\mathrm{M}\mathbf{x} = \mathbf{x}^T \mathrm{M}^T \mathrm{V}\Lambda\mathrm{V}^T \mathrm{M}\mathbf{x}$$

where V *is the orthonormal vector matrix which diagonalizes matrix* Q, *i.e.,* $\mathrm{Q} = \mathrm{V}\Lambda\mathrm{V}^T$.

Given the state transformation $\tilde{\mathbf{x}} = \mathrm{V}^T\mathrm{M}\mathbf{x}$, *then*

$$V(\tilde{\mathbf{x}}) = \tilde{\mathbf{x}}^T \Lambda \tilde{\mathbf{x}}$$

which shows that $\mathrm{M}^T\mathrm{Q}\mathrm{M}$ *is positive definite.*

Generalizations of theorem 5. Notice that with a similar procedure, if instead matrix M is not a full rank, matrix $\mathrm{M}^T\mathrm{Q}\mathrm{M}$ is positive semi-definite. Finally, if matrix M is not square ($\mathrm{M} \in \mathbb{R}^{n \times m}$) it can be shown that matrix $\mathrm{M}^T\mathrm{Q}\mathrm{M}$ is positive semi-definite. In both cases, in fact, the transformation $\tilde{\mathbf{x}} = \mathrm{V}^T\mathrm{M}\mathbf{x}$ is no longer an invertible transformation (also if $\mathrm{M} \in \mathbb{R}^{n \times m}$, then $\mathbf{x} \in \mathbb{R}^m$), so it can only be concluded that the product matrix is positive semi-definite.

Proof 3 *(Proof of the necessary part of Lyapunov's II criterion for linear systems). The proof of the necessary part of the theorem, that is if the system is asymptotically stable, then for every positive definite matrix* Q *there is a positive definite matrix* P *which is the solution of the Lyapunov equation (2.4), is a constructive demonstration introducing the integral solution of the Lyapunov equation.*

Remember that, if a linear time-invariant system is asymptotically stable, then the zero-input response tends to zero for any initial condition

$$\lim_{t\to+\infty} \mathbf{x}(t) = 0 \quad \forall \mathbf{x}_0$$

and, since $\mathbf{x}(t) = e^{\mathrm{A}t}\mathbf{x}(0)$, *also matrix* $e^{\mathrm{A}t}$ *tends to zero:*

$$\lim_{t\to+\infty} e^{\mathrm{A}t} = 0$$

Remember also that

$$[e^{\mathrm{A}t}]^T = e^{\mathrm{A}^T t}$$

This property is easily provable by considering the definition of an exponential matrix:

$$e^{\mathrm{A}t} = \mathrm{I} + \mathrm{A}t + \frac{\mathrm{A}^2t^2}{2!} + \frac{\mathrm{A}^3t^3}{3!} + \cdots \Rightarrow$$

$$(e^{\mathrm{A}t})^T = \mathrm{I} + \mathrm{A}^Tt + \frac{(\mathrm{A}^2)^Tt^2}{2!} + \frac{(\mathrm{A}^3)^Tt^3}{3!} + \cdots =$$

$$= \mathrm{I} + \mathrm{A}^Tt + \frac{(\mathrm{A}^T)^2t^2}{2!} + \frac{(\mathrm{A}^T)^3t^3}{3!} + \cdots = e^{\mathrm{A}^Tt}$$

The improper integral $\mathrm{P} = \int_0^\infty e^{\mathrm{A}^Tt}\mathrm{Q}e^{\mathrm{A}t}dt$ *is finite, since each integrating term tends to zero. Matrix* P, *because of the properties of theorem 5, is the integral of a positive definite matrix for each t and therefore is positive definite. Note that matrix* P *is also symmetrical (if* Q *is symmetrical).*

Given that $\mathrm{P} = \int_0^\infty e^{\mathrm{A}^Tt}\mathrm{Q}e^{\mathrm{A}t}dt$ *is a positive definite symmetrical matrix, the solution of Lyapunov equation (2.4) can be verified.*

By substituting P *into the Lyapunov equation we obtain:*

$$\mathrm{A}^T \int_0^\infty e^{\mathrm{A}^Tt}\mathrm{Q}e^{\mathrm{A}t}dt + \int_0^\infty e^{\mathrm{A}^Tt}\mathrm{Q}e^{\mathrm{A}t}dt\mathrm{A} =$$

$$= \int_0^\infty (\mathrm{A}^Te^{\mathrm{A}^Tt}\mathrm{Q}e^{\mathrm{A}t} + e^{\mathrm{A}^Tt}\mathrm{Q}e^{\mathrm{A}t}\mathrm{A})dt =$$

$$= \int_0^\infty \frac{d}{dt}(e^{\mathrm{A}^Tt}\mathrm{Q}e^{\mathrm{A}t})dt = \left[e^{\mathrm{A}^Tt}\mathrm{Q}e^{\mathrm{A}t}\right]_0^\infty = -Q$$

where we used the fact that

$$\lim_{t\to\infty} e^{\mathrm{A}t} = 0$$

and that $e^{\mathrm{A}0} = I$.

Note 1. The solution to Lyapunov's equation $AP + PA^T = -Q$ is $P = \int_0^\infty e^{At} Q e^{A^T t} dt$.

Note 2. To study the asymptotic stability of a linear time-invariant system from its characteristic polynomial, besides the well-known Routh criterion, the Hurwitz criterion can be used. It facilitates knowing when a given polynomial is a Hurwitz polynomial (i.e., when it has only roots with negative real part).

Theorem 6 (Hurwitz criterion) *Polynomial*

$$a(\lambda) = a_n \lambda^n + a_{n-1}\lambda^{n-1} + \ldots + a_1 \lambda + a_0$$

with $a_n > 0$ *is Hurwitz if all the principal minors of the so-called Hurwitz matrix are greater than zero:*

$$H = \begin{bmatrix} a_{n-1} & a_{n-3} & a_{n-5} & \cdots & 0 & 0 \\ a_n & a_{n-2} & a_{n-4} & \cdots & 0 & 0 \\ 0 & a_{n-1} & a_{n-3} & \cdots & 0 & 0 \\ 0 & a_n & a_{n-2} & \cdots & 0 & 0 \\ \vdots & \vdots & \vdots & \vdots & \vdots & \vdots \\ 0 & 0 & \cdots & \cdots & a_1 & 0 \\ 0 & 0 & \cdots & \cdots & a_2 & a_0 \end{bmatrix}$$

MATLAB® exercise 2.2

Consider the following polynomial which is obviously not a Hurwitz polynomial:

$$a(\lambda) = \lambda^4 + 2\lambda^3 + 5\lambda^2 - 10\lambda + 0.7$$

The Hurwitz matrix is given by:

$$H = \begin{bmatrix} 2 & -10 & 0 & 0 \\ 1 & 5 & 0.7 & 0 \\ 0 & 2 & -10 & 0 \\ 0 & 1 & 5 & 0.7 \end{bmatrix}$$

Calculating the principal minors produces: $H_1 = 2$; $H_2 = 20$; $H_3 = -202.8$ and $H_4 = -141.96$. As expected at least one of the minors is not positive ($H_3 < 0$ and $H_4 < 0$).

2.4 Lyapunov equations

In the previous section we saw that, if the linear time-invariant is asymptotically stable, the Lyapunov equation (2.4) has only one solution.

Let us consider now a generic Lyapunov equation:

$$AX + XB = -C \tag{2.5}$$

where A, B and C are matrices of $\mathbb{R}^{n\times n}$ (in the most general case, they are non-symmetrical), and X is a $n \times n$ unknown matrix.

Even here, the equation can be vectorized:

$$\mathrm{M}\begin{bmatrix} x_{11} \\ x_{12} \\ x_{13} \\ \vdots \\ x_{nn} \end{bmatrix} = -\begin{bmatrix} c_{11} \\ c_{12} \\ c_{13} \\ \vdots \\ c_{nn} \end{bmatrix}$$

where M is a suitable matrix of $\mathbb{R}^{n^2\times n^2}$ (if $\mathrm{B} = \mathrm{A}^T$ and C is a symmetrical matrix, the number of variables reduces to $k = \frac{n(n+1)}{2}$ and $\mathrm{M} \in \mathbb{R}^{k\times k}$).

The existence of a solution is related to the invertibility of matrix M. This matrix can be simply obtained with Kronecker algebra. Let's recall therefore some of the definitions and properties of Kronecker algebra.

The Kronecker product of two matrices A and B of $\mathbb{R}^{n\times n}$ is a matrix of $\mathbb{R}^{n^2\times n^2}$ defined by:

$$\mathrm{A}\otimes\mathrm{B} = \begin{bmatrix} a_{11}\mathrm{B} & a_{12}\mathrm{B} & \dots & a_{1n}\mathrm{B} \\ a_{21}\mathrm{B} & a_{22}\mathrm{B} & \dots & a_{2n}\mathrm{B} \\ \vdots & \vdots & & \vdots \\ a_{n1}\mathrm{B} & a_{n2}\mathrm{B} & \dots & a_{nn}\mathrm{B} \end{bmatrix}$$

The matrix's eigenvalues can be calculated from those of matrix A and B. Let us indicate with $\lambda_1, \lambda_2, \dots, \lambda_n$ the eigenvalues of A and with $\mu_1, \mu_2, \dots, \mu_n$ the eigenvalues of B. Then, the n^2 eigenvalues of $\mathrm{A}\otimes\mathrm{B}$ are:

$$\lambda_1\mu_1, \lambda_1\mu_2, \dots, \lambda_1\mu_n, \lambda_2\mu_1, \dots, \lambda_n\mu_n$$

It can be shown that the vectorization matrix is given by:

$$\mathrm{M} = \mathrm{A}\otimes\mathrm{I} + \mathrm{I}\otimes\mathrm{B}^T$$

with $\mathbf{I} \in \mathbb{R}^{n\times n}$. Even the eigenvalues of matrix M are simply linked to those of matrices A and B.

The eigenvalues of $\mathrm{M} = \mathrm{A}\otimes\mathrm{I} + \mathrm{I}\otimes\mathrm{B}^T$ are given by all the possible n^2 combinations of sums between an eigenvalue from A and one from B:

$$\lambda_1 + \mu_1, \lambda_1 + \mu_2, \dots, \lambda_1 + \mu_n, \lambda_2 + \mu_1, \dots, \lambda_n + \mu_n$$

M is invertible when there are no null eigenvalues:

$$\lambda_i + \mu_j \neq 0, \quad \forall i, j$$

This leads to whether a Lyapunov equation has a solution or not. Formally, the condition above can be expressed with the following theorem.

Theorem 7 *Lyapunov equation*

$$\mathrm{AX} + \mathrm{XB} = -\mathrm{C}$$

where $\mathrm{A} \in \mathbb{R}^{n \times n}$*,* $\mathrm{B} \in \mathbb{R}^{n \times n}$ *and* $\mathrm{C} \in \mathbb{R}^{n \times n}$ *produces only one solution if and only if*

$$\lambda_i + \mu_j \neq 0, \quad \forall i, j$$

where $\lambda_1, \ldots, \lambda_n$ *are the eigenvalues of* A *and* $\mu_1, \ldots, \mu_n$ *are those of* B.

An immediate consequence of this theorem is that the Lyapunov equation $\mathrm{A}^T\mathrm{P} + \mathrm{PA} = -\mathrm{Q}$ has only one solution if A has no eigenvalues on the imaginary axis and if there are no pairs of real eigenvalues with equal module and opposite sign (for example, $\lambda_1 = -1$ and $\lambda_2 = +1$), all of which conditions are verified by the hypothesis of asymptotic stability.

Note. We have seen that vectorizing the Lyapunov equation requires inverting a matrix of size $n^2 \times n^2$ (or a matrix of size $\frac{n(n+1)}{2} \times \frac{n(n+1)}{2}$). Even for a system of order $n = 10$ it means a matrix of ten thousand items. This method is more laborious to calculate than other iterative methods used routinely for solving Lyapunov equations. Below is the sketch of an algorithm based on a Schur decomposition for solving Lyapunov equations of type $\mathrm{A}^T\mathrm{P}+\mathrm{PA} = -\mathrm{Q}$. A Schur decomposition able to obtain $\mathrm{A} = \mathrm{U}\bar{\mathrm{A}}\mathrm{U}^T$ with $\mathrm{U}^T\mathrm{U} = \mathrm{I}$ and higher triangular matrix $\bar{\mathrm{A}}$ is considered. At this point, all the Lyapunov equation terms are multiplied by U^T left and U right. Thus, $\bar{\mathrm{A}}^T\bar{\mathrm{P}} + \bar{\mathrm{P}}\bar{\mathrm{A}} = -\bar{\mathrm{Q}}$ is obtained with $\bar{\mathrm{Q}} = \mathrm{U}^T\mathrm{QU}$ and $\bar{\mathrm{P}} = \mathrm{U}^T\mathrm{PU}$. Here, triangular matrix A can be used to iteratively solve the equation and then to find the solution P from $\mathrm{P} = \mathrm{U}\bar{\mathrm{P}}\mathrm{U}^T$.

MATLAB® exercise 2.3

The aim of this MATLAB exercise is to familiarize some MATLAB commands for resolving control issues. It is always advisable to refer to the "help" command to learn about how each function works.

Let's consider a linear time-invariant system with state matrix

$$\mathrm{A} = \begin{bmatrix} -1 & 0 & 1 & 0 \\ 2 & 0 & 3 & 5 \\ 1 & 10 & 0.5 & 0 \\ -2 & 3 & -5 & -10 \end{bmatrix}$$

and then calculate matrix P of Lyapunov equation (2.4) with $\mathrm{Q} = \mathrm{I}$ by vectorization. To do this, matrix $\mathrm{M} = \mathrm{I} \otimes \mathrm{A}^T + \mathrm{A}^T \otimes \mathrm{I}$ with $\mathrm{C} = \mathrm{I}$ is constructed.

1. Define matrix A:

```
>> A=[-1 0 1 0; 2 0 3 5; 1 10 0.5 0; -2 3 -5 -10];
```

and identity matrix:

```
>> MatriceI=eye(4);
```

2. Construct the vector $\begin{bmatrix} c_{11} \\ c_{12} \\ c_{13} \\ \vdots \\ c_{nn} \end{bmatrix}$

```
>> c=MatriceI(:)
```

3. Construct matrix M

```
>> M=kron(eye(4),A')+kron(A',eye(4))
```

4. Find the solution vector by inverting matrix M

```
>> Psol=-inv(M)*c;
```

and obtain matrix P with an appropriate size (4×4)

```
>> P=reshape(Psol,4,4)
```

One obtains $P = \begin{bmatrix} 0.4727 & 0.2383 & -0.1661 & 0.1690 \\ 0.2383 & 1.7516 & -0.3188 & 0.8961 \\ -0.1661 & -0.3188 & -0.0466 & -0.1292 \\ 0.1690 & 0.8961 & -0.1292 & 0.4980 \end{bmatrix}$.

5. Verify that matrix P solves the problem

```
>> A'*P+P*A
```

MATLAB can solve Lyapunov equations more efficiently with the `lyap` command. In this case

```
>> P=lyap(A',eye(4))
```

In conclusion, using command `eig(P)` to calculate the eigenvalues of P, it can be verified that P is not a positive definite matrix and that the system is not asymptotically stable which is the case after calculating the eigenvalues of A ($\lambda_1 = -13.0222$, $\lambda_2 = 4.7681$, $\lambda_3 = -1.1229 + 0.8902i$, $\lambda_4 = -1.1229 - 0.8902i$).

2.5 Stability with uncertainty

Concluding this chapter is a brief digression on stability with uncertainty. We will confine ourselves to presenting one key finding for the stability of systems with parametric uncertainty.

Let $D(s)$ be the characteristic polynomial of a linear time-invariant system. Suppose the uncertainty of the system can be expressed in terms of variation ranges of coefficients of polynomial $D(s) = a_n s^n + a_{n-1}s^{n-1} + \ldots + a_1 s + a_0$:

$$\begin{array}{c} a_0^m \leq a_0 \leq a_0^M \\ \ldots \\ a_n^m \leq a_n \leq a_n^M \end{array} \tag{2.6}$$

So, the hypothesis is that parametric uncertainty can be characterized by assigning maximum and minimum values that the various coefficients of $D(s)$ might assume. Under this hypothesis, stability with uncertainty can be studied very effectively and quickly thanks to the Kharitonov criterion.

Theorem 8 (Kharitonov criterion) *Let* $D(s) = a_n s^n + a_{n-1}s^{n-1} + a_{n-2}s^{n-2} + a_{n-3}s^{n-3} + \dots$ *be the characteristic polynomial of a linear time-invariant system with coefficients such that* $a_i^m \leq a_i \leq a_i^M$ *for* $i = 0, 1, \dots, n$, *if and only if the four polynomials*

$$\begin{array}{l} D_1(s) = a_n^m s^n + a_{n-1}^m s^{n-1} + a_{n-2}^M s^{n-2} + a_{n-3}^M s^{n-3} + \dots \\ D_2(s) = a_n^M s^n + a_{n-1}^M s^{n-1} + a_{n-2}^m s^{n-2} + a_{n-3}^m s^{n-3} + \dots \\ D_3(s) = a_n^m s^n + a_{n-1}^M s^{n-1} + a_{n-2}^M s^{n-2} + a_{n-3}^m s^{n-3} + \dots \\ D_4(s) = a_n^M s^n + a_{n-1}^m s^{n-1} + a_{n-2}^m s^{n-2} + a_{n-3}^M s^{n-3} + \dots \end{array} \tag{2.7}$$

are stable, then the polynomial $D(s)$ *is stable for any parameter whatsoever.*

The Kharitonov criterion facilitates the study of stability in a parametric polynomial (aided for example by the Routh criterion). Quite independently of the order of the system and therefore of the number of parameters, this criterion can calculate the stability of an infinite number of systems provided they have limited polynomial coefficients.

MATLAB® exercise 2.4

Let us consider the system with characteristic polynomial

$$D(s) = a_5 s^5 + a_4 s^4 + a_3 s^3 + a_2 s^2 + a_1 s + a_0 \tag{2.8}$$

with uncertain parameters:

$$\begin{array}{c} 1 \leq a_5 \leq 2 \\ 3 \leq a_4 \leq 5 \\ 5 \leq a_3 \leq 7 \\ \frac{4}{3} \leq a_2 \leq \frac{5}{2} \\ \frac{1}{2} \leq a_1 \leq \frac{3}{4} \\ 1 \leq a_0 \leq 2 \end{array}$$

To apply the Kharitonov criterion the following characteristic polynomials have to be considered:

$$\begin{array}{l} D_1(s) = s^5 + 3s^4 + 7s^3 + \frac{5}{2}s^2 + \frac{1}{2}s + 1 \\ D_2(s) = 2s^5 + 5s^4 + 5s^3 + \frac{4}{3}s^2 + \frac{3}{4}s + 2 \\ D_3(s) = s^5 + 5s^4 + 7s^3 + \frac{4}{3}s^2 + \frac{3}{4}s + 2 \\ D_4(s) = 2s^5 + 3s^4 + 5s^3 + \frac{5}{2}s^2 + \frac{3}{4}s + 1 \end{array} \tag{2.9}$$

These can be defined in MATLAB as follows

```
>> D1=[1 3 7 5/2 1/2 1]
>> D2=[2 5 5 4/3 3/4 2]
>> D3=[1 5 7 4/3 1/2 2]
>> D4=[2 3 5 5/2 3/4 1]
```

Their roots are then calculated:

```
>> roots(D1)
>> roots(D2)
>> roots(D3)
>> roots(D4)
```

obtaining the following roots:

- $D_1(s)$: $s_{1,2} = -1.3073 \pm j2.0671$; $s_3 = -0.7059$; $s_{4,5} = 0.1603 \pm j0.4595$;

- $D_2(s)$: $s_{1,2} = -1.0224 \pm j0.9384$; $s_3 = -1.1541$; $s_{4,5} = 0.3494 \pm j0.5726$;
- $D_3(s)$: $s_{1,2} = -2.3261 \pm j0.4235$; $s_3 = -0.9566$; $s_{4,5} = 0.3044 \pm j0.5304$;
- $D_4(s)$: $s_{1,2} = -0.5107 \pm j1.3009$; $s_3 = -0.8031$; $s_{4,5} = 0.1623 \pm j0.5408$.

From the inspection of the roots of $D_1(s)$, $D_2(s)$, $D_3(s)$ and $D_4(s)$, it can be concluded that the system with characteristic polynomial (2.8) is not stable.

MATLAB® exercise 2.5

Determine the locus of the closed-loop eigenvalues for $k = 10$, $k = 30$ and $k = 120$ for the system $P(s, q) = \frac{1}{s(s^2+(8+q_1)s+(20+q_2))}$ with $-2 \leq q_1 \leq 2$ and $-4 \leq q_2 \leq 4$.

Solution

First, determine the characteristic closed-loop polynomial:

$$p(s, q, k) = s^3 + (8 + q1)s^2 + (20 + q2)s + k$$

This is a parametric polynomial that can be re-written as:

$$p(s, q, k) = s^3 + As^2 + Bs + k$$

with $A = 8 + q1$ and $B = 20 + q2$. The two parameters in the characteristic polynomial vary in $[6, 10]$ and in $[16, 24]$, respectively. At this point, apply the Kharithonov criterion to know if, by fixing k, the asymptotic stability is guaranteed in the parametric uncertainty interval. Indicated as $A_{min} = 6$, $A_{max} = 10$, $B_{min} = 16$ and $B_{max} = 24$, we need to study the roots of the four polynomials:

$$p(s, q, k) = s^3 + A_{min}s^2 + B_{max}s + k$$

$$p(s, q, k) = s^3 + A_{max}s^2 + B_{min}s + k$$

$$p(s, q, k) = s^3 + A_{min}s^2 + B_{min}s + k$$

$$p(s, q, k) = s^3 + A_{max}s^2 + B_{max}s + k$$

To do this, the following commands in MATLAB are used:

```
>> k=12;
>> Amin=6; Amax=10; Bmin=16; Bmax=24;
>> A=Amin; B=Bmax; roots([1 A B k])
>> A=Amax; B=Bmin; roots([1 A B k])
>> A=Amin; B=Bmin; roots([1 A B k])
>> A=Amax; B=Bmax; roots([1 A B k])
```

While for $k = 10$ and $k = 30$ we obtain polynomials having roots with negative real part, for $k = 120$ the third polynomial has two roots with positive real part. After this preliminary analysis, the locus can be built with a Monte Carlo simulation, assigning the parameter values one by one, calculating their roots and plotting them on a graph. By fixing k, we can use the following commands in MATLAB:

```
>> plot(0,0,'m.'); hold on;
>> for A=[Amin:0.1:Amax]
>>     for B=[Bmin:0.1:Bmax]
>>         p=roots([1 A B k]);
>>         plot(real(p), imag(p),'x')
>>     end
>> end
```

Figure 2.2 shows the plots for four different values of k. Notice that for $k = 120$ some roots lie on the right half plane.

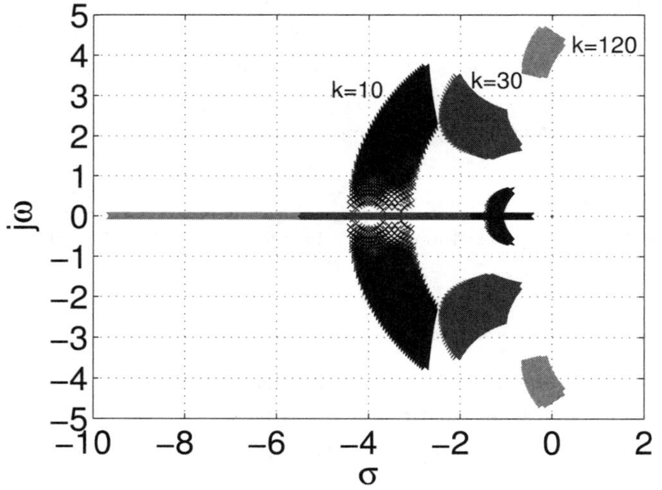

FIGURE 2.2
Locus of the closed-loop eigenvalues for exercise 2.5.

2.6 Exercises

1. Apply the vectorization method to solve the Lyapunov equation:

$$\mathrm{A}^T\mathrm{P} + \mathrm{PA} = -\mathrm{I}$$

with

$$\mathrm{A} = \begin{bmatrix} 0 & 1 & -1 \\ 2 & -5 & -1 \\ 3 & 1 & -2 \end{bmatrix}$$

2. Given the nonlinear system $\dot{x} = x^3 - 8x^2 + 17x + u$ calculate the equilibrium points for $u = -10$ and study their stability.
3. Given system $G(s) = \frac{1}{s}e^{-sT}$ determine the set of T values which provide a stable closed-loop system.
4. Study the stability of system with transfer function

$$G(s) = \frac{s^2 + 2s + 2}{s^4 + a_1 s^3 + a_2 s^2 + a_3 s + a_4}$$

with $1 \leq a_1 \leq 3$, $4 \leq a_2 \leq 7$, $1 \leq a_3 \leq 2$, $0.5 \leq a_4 \leq 2$.

5. Given the polynomial $p(s,a) = s^4 + 5s^3 + 8s^2 + 8s + 3$ with $a = \begin{bmatrix} 3 & 8 & 8 & 5 \end{bmatrix}$, find $p(s,b)$ with $b = \begin{bmatrix} (b_0^-, b_0^+) & \dots & (b_3^-, b_3^+) \end{bmatrix}$ so that the polynomial class $p(s,b)$ is Hurwitz.

6. Study the stability of the system $G(s) = \frac{s^2+3s+2}{s^4+q_1s^3+5s^2+q_2s+q_3}$ with parameters $q_1 \in [1,3]$, $q_2 \in [5,10]$, $q_3 \in [2,18]$.

3

Kalman canonical decomposition

CONTENTS

This chapter describes Kalman canonical decomposition, which highlights the state variables that do not affect the input/output properties of the system, but which nevertheless may be very important.

This quite general discussion refers to continuous-time systems, so there is no distinction between the controllability properties and the reachability of the system.

3.1 Introduction

Given a continuous-time linear time-invariant system

$$\begin{aligned} \dot{\mathbf{x}} &= \mathrm{A}\mathbf{x} + \mathrm{B}\mathbf{u} \\ \mathbf{y} &= \mathrm{C}\mathbf{x} \end{aligned} \tag{3.1}$$

recall that the transfer matrix of the system is given by $G(s) = \mathrm{C}(s\mathrm{I}-\mathrm{A})^{-1}\mathrm{B}$, the controllability matrix $\mathrm{M}_c = \begin{bmatrix} \mathrm{B} & \mathrm{AB} & \dots & \mathrm{A}^{n-1}\mathrm{B} \end{bmatrix}$, the observability matrix $M_o = \begin{bmatrix} C \\ CA \\ \vdots \\ CA^{n-1} \end{bmatrix}$ and the Hankel matrix

$$H = M_o M_c = \begin{bmatrix} CB & CAB & \dots & CA^{n-1}B \\ CAB & CA^2B & \dots & \dots \\ \vdots & \vdots & \vdots & \vdots \\ CA^{n-1}B & \dots & \dots & \dots \end{bmatrix}.$$

Remember also that if the controllability matrix has full rank, that is equal to n (i.e., the system order), then the system is completely controllable. If the observability matrix has full rank, that is equal to n, then the system is completely observable. If the Hankel matrix has rank n, then the system is completely controllable and observable.

Now consider an invertible linear transformation $\tilde{\mathbf{x}} = T^{-1}\mathbf{x}$. The state matrices of the new reference system are related to the original state matrices through the relations:

$$\begin{aligned} \tilde{A} &= T^{-1}AT \\ \tilde{B} &= T^{-1}B \\ \tilde{C} &= CT \end{aligned} \tag{3.2}$$

and, as we know, the transfer matrix is a system invariant: it does not change even as the reference system changes: $\tilde{G}(s) = \tilde{C}(sI - \tilde{A})^{-1}\tilde{B} = C(sI - A)^{-1}B = G(s)$.

The controllability and observability matrices vary as the reference system varies. In particular, they are given by:

$$\begin{aligned} \tilde{M}_c &= T^{-1}M_c \\ \tilde{M}_o &= M_o T \end{aligned} \tag{3.3}$$

Since matrix T is invertible and therefore the rank of $\tilde{M}_c$ and $\tilde{M}_o$ correspond to the rank of M_c and of M_o, from the relations (3.3) we deduce the well-known property that controllability and observability are *structural properties* of the system.

From relations (3.3) note that $\tilde{M}_o\tilde{M}_c = M_o M_c$ so the Hankel matrix is also an invariant of the system. Remember also that a system is minimal when it is completely controllable and observable, that is when the rank of the Hankel matrix equals n.

The number of state variables which characterize a minimal system is the minimum number of state variables needed to express all the input/output relations of the system. A system is minimal when its minimal form has order n, the system order.

For example, the system with transfer function $G(s) = \frac{s+1}{(s+1)(s+2)}$ is not minimal. However, we could say this is a "lucky" case, because the underlying dynamics (i.e., that which does not influence the input/output relations) is stable. Generally, even hidden state variables are very significant, and it is the stability associated to the hidden dynamics that is the most significant property to be taken into consideration.

Consider also another example, represented by system $G(s) = \frac{s+0.98}{(s+1)(s+2)}$. There is not an exact simplification between the pole and zero, but clearly the pole and zero are very close and in the presence of uncertainty a simplification can occur. Conversely, a seemingly exact simplification is less precise in cases of uncertainty. The issue of determining a system minimal form is thus closely related to structural uncertainty.

In this chapter we will examine the Kalman decomposition which takes into consideration when the simplifications between the poles and zeros of a system are exact, but in the next chapters we will raise the issue of determining a system approximation that takes into account the most significant dynamic for the input/output relation and that is the most robust against structural uncertainties.

3.2 Controllability canonical partitioning

Consider system (3.1) and suppose it is not completely controllable. This means there are internal state variables whose value, starting from a given initial condition and acting through the inputs, cannot be set to a value arbitrarily fixed in the system state-space. An example of an uncontrollable system is shown in Figure 3.1. Let $x_1(t)$ and $x_2(t)$ indicate the voltage across capacitors C_1 and C_2, respectively. Suppose $C_1 \neq C_2$, since the two capacitors are connected in series, the charge held by them is equal, then $\frac{C_1}{C_2} = \frac{x_1}{x_2}$. Therefore the system is not controllable, and for example the state $x_1 = x_2 = 5V$ is not attainable by the system.

Canonical decomposition for controllability is a state representation which highlights the division of state variables into controllable state variables and uncontrollable state variables. Canonical decomposition thus emphasizes the existence of a controllable part and an uncontrollable part of the system. $\mathbf{z}$ indicates the state variables of the new reference system. They can be partitioned into two subsets: $\mathbf{z}_r$ are controllable variables and $\mathbf{z}_{nr}$ the uncontrollable variables. The system equations in Kalman canonical form for controllability can be expressed as follows:

$$\begin{aligned} \dot{\mathbf{z}_r} &= \mathrm{A}_1\mathbf{z}_r + \mathrm{A}_{12}\mathbf{z}_{nr} + \mathrm{B}_1\mathbf{u} \\ \dot{\mathbf{z}_{nr}} &= \mathrm{A}_2\mathbf{z}_{nr} \end{aligned} \tag{3.4}$$

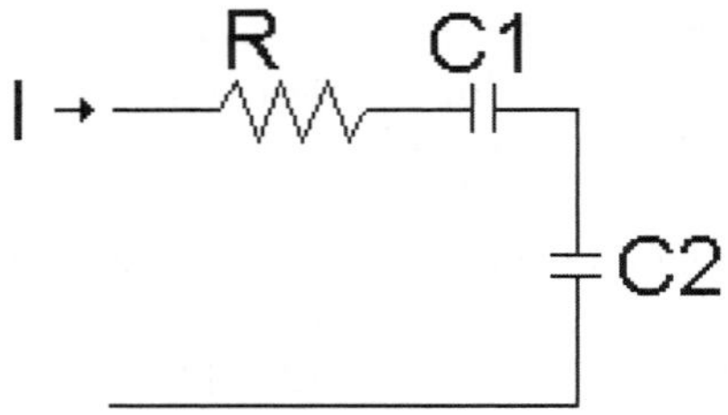

FIGURE 3.1
Example of a non-controllable system.

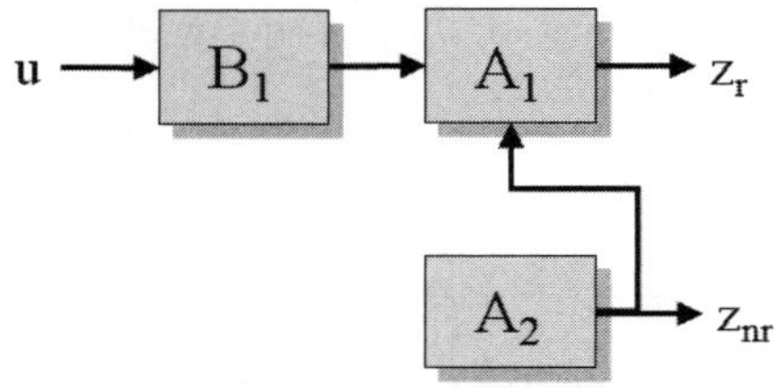

FIGURE 3.2
Canonical decomposition for controllability.

The state matrices are thus block-partitioned matrices: $\tilde{A} = \begin{bmatrix} A_1 & A_{12} \\ 0 & A_2 \end{bmatrix}$ and $\tilde{B} = \begin{bmatrix} B_1 \\ 0 \end{bmatrix}$.

The diagram of this model is shown in Figure 3.2. Note that it is not possible to act through the inputs on the variables $\mathbf{z}_{nr}$ directly given that $B_2 = 0$, nor indirectly through the other state variables $\mathbf{z}_r$ given that $A_{21} = 0$.

Note also that, given the structure of matrix $\tilde{A}$, the system eigenvalues are given by the union of those of A_1 and those of A_2. If A_1 has unstable eigenvalues, those acting through a control law can be moved to the left-hand half of the complex plane. Instead, if A_2 has unstable eigenvalues, it is impossible to act on them. This second case is the most serious, because the system cannot be stabilized by feedback, in which case the system is known as non-stabilizable. By contrast, a system for which all unstable eigenvalues belong to the controllable part is known as stabilizable.

Having explained the characteristics of canonical decomposition for controllability, let's see how to calculate matrix T which allows us to switch from the original form to the particular structure of canonical decomposition.

Consider the controllability matrix M_c. It is rank deficient. Note however, that the linearly independent columns of this matrix define the subspace of controllability of the system. The subspace of uncontrollability can be defined as orthogonal to that of controllability. Matrix T is constructed from the linearly independent columns of M_c. In particular, $T = \begin{bmatrix} T_1 & T_2 \end{bmatrix}$ where T_1 is the matrix formed by the linearly independent columns of M_c and T_2 by the column-vectors orthogonal to those linearly independent of M_c. Applying this transformation matrix to the original system we obtain:

$$\tilde{A} = T^{-1}AT = \begin{bmatrix} U_1 \\ U_2 \end{bmatrix} A \begin{bmatrix} T_1 & T_2 \end{bmatrix} = \begin{bmatrix} U_1AT_1 & U_1AT_2 \\ U_2AT_1 & U_2AT_2 \end{bmatrix}$$

but, since the columns of U_2 are orthogonal to the linearly independent columns of A, then $U_2AT_1 = 0$ and

$$\tilde{A} = \begin{bmatrix} U_1AT_1 & U_1AT_2 \\ 0 & U_2AT_2 \end{bmatrix}$$

In the same way we obtain that

$$\tilde{B} = T^{-1}B = \begin{bmatrix} U_1B_1 \\ 0 \end{bmatrix}$$

3.3 Observability canonical partitioning

There is a dual decomposition for observability which can identify the observable and unobservable parts of a system. The reference diagram is shown in Figure 3.3, while the system model in this particular state-space representation is expressed by the following equations:

$$\begin{aligned} \dot{\mathbf{z}}_o &= A_1\mathbf{z}_o + B_1\mathbf{u} \\ \dot{\mathbf{z}}_{no} &= A_{21}\mathbf{z}_o + A_2\mathbf{z}_{no} + B_2\mathbf{u} \\ \mathbf{y} &= C_1\mathbf{z}_o \end{aligned} \tag{3.5}$$

Again the state matrices are partitioned in blocks: $\tilde{A} = \begin{bmatrix} A_1 & 0 \\ A_{21} & A_2 \end{bmatrix}$, $\tilde{B} = \begin{bmatrix} B_1 \\ B_2 \end{bmatrix}$ and $\tilde{C} = \begin{bmatrix} C_1 & 0 \end{bmatrix}$.

The model can be derived analogously to the canonical decomposition for controllability. The unobservable variables z_{no} cannot be reconstructed from input/output measurements so there cannot be any direct link between these variables and the output. Because $C_2 = 0$, the unobservable variables do not directly affect system output. Similarly, since the unobservable variables cannot be reconstructed from the output information, even indirectly, they cannot influence the dynamics of the observable variables. So, $A_{12} = 0$.

The transformation matrix from the original reference system to the canonical decomposition for observability can be found with a dual procedure of the previous one.

This time, let's consider the linearly independent rows of observability matrix M_o to construct matrix U_1. Suppose there are n_o linearly independent rows. Consider the subspace of size $n - n_o$ orthogonal to the defined one and determine a subspace basis to construct matrix U_2. So, we have $T^{-1} = \begin{bmatrix} U_1 \\ U_2 \end{bmatrix}$. Applying this transformation matrix to the original system we have

$$\tilde{A} = \begin{bmatrix} U_1AT_1 & U_1AT_2 \\ U_2AT_1 & U_2AT_2 \end{bmatrix} = \begin{bmatrix} U_1AT_1 & 0 \\ U_2AT_1 & U_2AT_2 \end{bmatrix}$$

and

$$\tilde{C} = \begin{bmatrix} CT_1 & CT_2 \end{bmatrix} = \begin{bmatrix} CT_1 & 0 \end{bmatrix}$$

Again the system eigenvalues are the union of the eigenvalues of A_1 and those of A_2. If the system is unstable and controllable, but not observable, we need to know to which part belong the unstable modes of the system. If they are located in the unobservable part, these modes cannot be observed and so they cannot be controlled. However, if they are located in the observable part of the system, they can be reconstructed and then controlled.

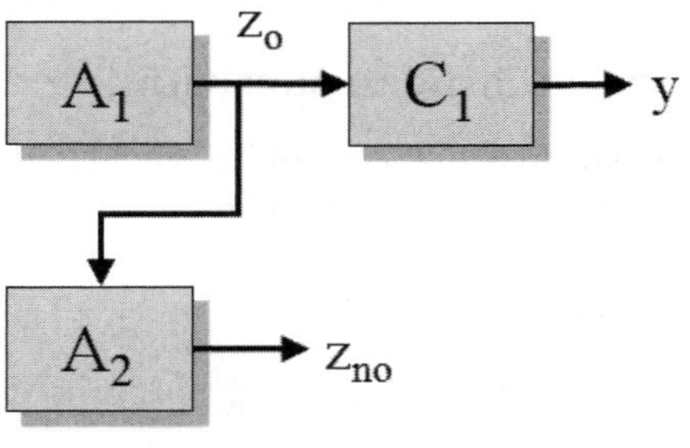

FIGURE 3.3
Observability canonical decomposition.

3.4 General partitioning

The two decompositions in the paragraphs above are elementary decompositions which can de-couple a system into a controllable part and an uncontrollable part or into an observable part and an unobservable part. There is a more general decomposition which can split a system into four parts: a controllable and unobservable part (part A); a controllable and observable part (part B); an uncontrollable and unobservable part (part C); an uncontrollable and observable part (part D). Clearly, this decomposition can be applied to the most general case in which the system is not completely controllable nor observable.

The state variables corresponding to part A of the system are $\mathbf{z}_{r,no}$; part B $\mathbf{z}_{r,o}$; part C $\mathbf{z}_{nr,no}$; and to part D $\mathbf{z}_{nr,o}$. The block diagram of the Kalman decomposition is shown in Figure 3.4 and the corresponding state-space equations are characterized by the following matrices:

$$\tilde{\mathrm{A}} = \begin{bmatrix} \mathrm{A}_A & \mathrm{A}_{AB} & \mathrm{A}_{AC} & \mathrm{A}_{AD} \\ 0 & \mathrm{A}_B & 0 & \mathrm{A}_{BD} \\ 0 & 0 & \mathrm{A}_C & \mathrm{A}_{AD} \\ 0 & 0 & 0 & \mathrm{A}_D \end{bmatrix}; \tilde{\mathrm{B}} = \begin{bmatrix} \mathrm{B}_A \\ \mathrm{B}_B \\ 0 \\ 0 \end{bmatrix}; \tilde{\mathrm{C}} = \begin{bmatrix} 0 & \mathrm{C}_B & 0 & \mathrm{C}_D \end{bmatrix}$$

Note that matrix $\tilde{\mathrm{A}}$ is a triangular-block matrix. The zero blocks reflect no direct link between two parts of the system. For example, since the controllable and the unobservable part of the system cannot influence the controllable and observable part of the system (otherwise it would lose its property of being unobservable), block A_{BA} is zero. Analogous considerations apply to all the possible connections from any block in Figure 3.4 to any other block below. Any of these links would violate one of the hypotheses underlying the structural properties of the various parts of the system, so they are impossible. Conversely, the upward links between blocks are permitted except any link between the uncontrollable and unobservable part (part C) and the controllable and observable part (part B). In fact, if there was such a link, part C would no longer be unobservable. For this reason $\mathrm{A}_{BC} = 0$.

Now let's analyze how to find transition matrix T. Its construction requires defining four eigenvector groups which form the columns of matrix T and the bases of four subspaces which will be defined below, after having recalled some preliminary notions.

A subspace is A-invariant if, when any subspace vector is multiplied by matrix A, it still belongs to the subspace.

The reachability subspace and unobservability subspace are A-invariant subspaces. The first subspace constructed to define matrix T is given by the intersection of the reachability subspace X_r and the unobservability subspace X_{no}:

$$\mathrm{X}_A = \mathrm{X}_r \bigcap \mathrm{X}_{no} \tag{3.6}$$

This is an A-invariant subspace, since it is intersection of two A-invariant subspaces.

From X_A we then define X_B, so that the direct sum between X_A and X_B is the reachability subspace:

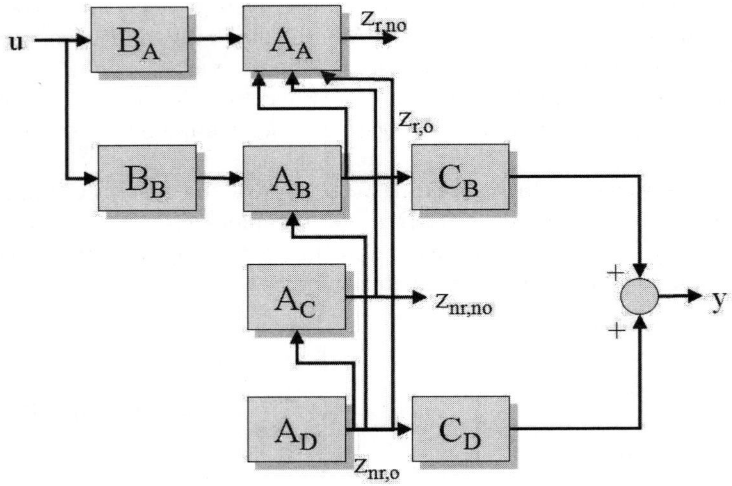

FIGURE 3.4
Canonical decomposition of a linear system.

$$X_r = X_A \oplus X_B \tag{3.7}$$

Similarly, X_C is defined so that the direct sum of X_A and X_C is the unobservability subspace:

$$X_{no} = X_A \oplus X_C \tag{3.8}$$

Finally, we construct X_D so as to obtain the entire state-space as a direct sum of X_r and X_{no} and of X_D.

$$X = (X_r + X_{no}) \oplus X_D \tag{3.9}$$

Since

$$X_A \oplus X_B \oplus X_C = (X_A \oplus X_B) + (X_A \oplus X_C) = X_r + X_{no}$$

then

$$X = X_A \oplus X_B \oplus X_C \oplus X_D$$

It follows then, that any state vector of the system can be represented uniquely as the sum of four vectors which belong to the spaces defined above. So, the state-space is decomposed into the direct sum of the four subspaces defined as X_A, X_B, X_C and X_D. From these subspaces the columns of matrix T can be generated. However, these subspaces are defined quite arbitrarily which

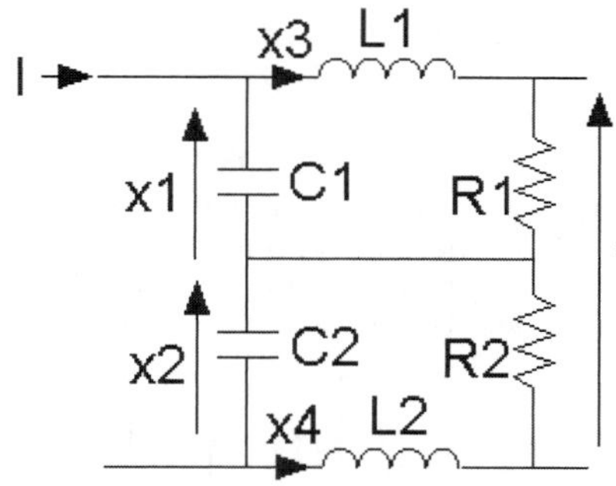

FIGURE 3.5
Example of electrical circuit illustrating Kalman decomposition.

can be prevented by imposing additional restrictions to them. In particular, we can make X_B and X_C orthogonal to X_A and that X_D is orthogonal to $(X_r + X_{no})$, thus obtaining new subspaces $\bar{X}_B$, $\bar{X}_C$ and $\bar{X}_D$. With these additional restrictions, subspaces X_A, $\bar{X}_B$, $\bar{X}_C$ and $\bar{X}_D$ can unequivocally be defined:

$$\begin{array}{l} X_A = X_r \bigcap X_{no} \\ \bar{X}_B = X_r \bigcap (X_{nr} + X_o) \\ \bar{X}_C = X_{no} \bigcap (X_{nr} + X_o) \\ \bar{X}_D == X_{nr} \bigcap X_o \end{array} \tag{3.10}$$

Once these subspaces are defined, we can obtain the transition matrix T. The columns of this matrix are provided by the bases of the generated subspaces. The procedure for calculating matrix T is described in detail in the following example.

Example 3.1

Consider the electrical circuit shown in Figure 3.5 and suppose that all the circuit components have normalized values: $L_1 = 1$, $L_2 = 1$, $C_1 = 1$, $C_2 = 1$, $R_1 = 1$ and $R_2 = 1$. The state variables of this system are the two voltages on the capacitors shown in the figure, $x_1(t)$ and $x_2(t)$, and the two currents in the inductors, $x_3(t)$ and $x_4(t)$. Given these state variables, the system is described by the following matrices:

$$A = \begin{bmatrix} 0 & 0 & -1 & 0 \\ 0 & 0 & 0 & 1 \\ 1 & 0 & -1 & 0 \\ 0 & -1 & 0 & -1 \end{bmatrix}; B = \begin{bmatrix} 1 \\ 1 \\ 0 \\ 0 \end{bmatrix}; C = \begin{bmatrix} 0 & 0 & 1 & -1 \end{bmatrix}$$

Let's first calculate the observability and controllability matrices:

$$M_o^T = \begin{bmatrix} 0 & 1 & -1 & 0 \\ 0 & 1 & -1 & 0 \\ 1 & -1 & 0 & 1 \\ -1 & 1 & 0 & 1 \end{bmatrix} \tag{3.11}$$

$$M_c = \begin{bmatrix} 1 & 0 & -1 & 1 \\ 1 & 0 & -1 & 1 \\ 0 & 1 & -1 & 0 \\ 0 & -1 & 1 & 0 \end{bmatrix} \tag{3.12}$$

Both matrices have a rank of two. A basis for the observability subspace X_o can be established from the observability matrix, taking for example the first and third columns (linearly independent). The unobservability subspace X_{no} is constructed orthogonal to the observability subspace X_o and is generated by the vectors:

$$V_1 = \begin{bmatrix} 1 \\ -1 \\ 0 \\ 0 \end{bmatrix}; V_2 = \begin{bmatrix} 0 \\ 0 \\ 1 \\ 1 \end{bmatrix}$$

The reachability subspace X_r is identified by the linearly independent columns of M_c. Take for example the first two columns of M_c. The unreachability subspace is determined by considering the subspace orthogonal to that identified by the first two columns of M_c.
Note how in this case $X_o = X_r$. So $X_r = X_{no}^{\perp}$ and $X_r \perp X_{no}$.
As the reachability subspace is orthogonal to the unobservability subspace, their intersection is the empty set. The intersection of these subspaces (see formula (3.6)) is X_A. So $X_A = \{\emptyset\}$.
Moreover, since $X_D = X_{nr} \bigcap X_o$, also $X_D = \{\emptyset\}$.
Finally, applying equation (3.10), $\bar{X}_B$ and $\bar{X}_C$ can be found:

$$\bar{X}_B = X_r \bigcap (X_{nr} + X_r) = X_r$$

$$\bar{X}_C = X_{no} \bigcap (X_{nr} + X_r) = X_{no}$$

From this, it follows that the transition matrix can be defined by taking the first two columns of matrix M_c and the vectors that form the basis of X_{no} (V_1 and V_2):

$$T = \begin{bmatrix} 1 & 0 & 1 & 0 \\ 1 & 0 & -1 & 0 \\ 0 & 1 & 0 & 1 \\ 0 & -1 & 0 & 1 \end{bmatrix} \tag{3.13}$$

At this point, the state matrices $\tilde{A}$, $\tilde{B}$ and $\tilde{C}$ in the state-space representation of the Kalman decomposition can be calculated:

$$\tilde{A} = T^{-1}AT = \begin{bmatrix} 0 & -1 & 0 & 0 \\ 1 & -1 & 0 & 0 \\ 0 & 0 & 0 & -1 \\ 0 & 0 & 1 & -1 \end{bmatrix}$$

$$\tilde{B} = T^{-1}B = \begin{bmatrix} 1 \\ 0 \\ 0 \\ 0 \end{bmatrix}$$

$$\tilde{C} = CT = \begin{bmatrix} 0 & 2 & 0 & 0 \end{bmatrix}$$

As you can see, part B of the system (i.e., the controllable and observable part) is second order and so also part C (the uncontrollable and unobservable part), whereas there is no controllable and unobservable part nor is there an observable and uncontrollable part. The transfer function of the system is therefore second order.

The next MATLAB® exercise shows another example of Kalman decomposition and how to obtain it by using MATLAB®.

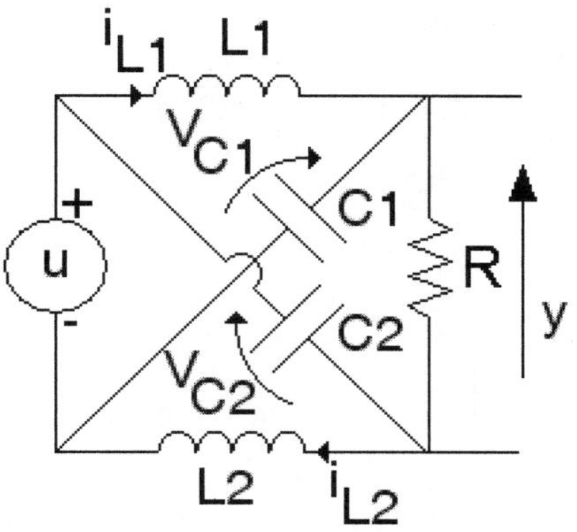

FIGURE 3.6
An *all-pass* electrical circuit.

MATLAB® exercise 3.1

Consider the circuit shown in Figure 3.6 (the reader is referred to the book of Wiener for more deep discussion on this circuit). Given $L_1 = L_2 = L$, $C_1 = C_2 = C$ and $R = \sqrt{\frac{L}{C}}$, it has the following transfer function:

$$\frac{Y(s)}{U(s)} = \frac{1 - s\sqrt{LC}}{1 + s\sqrt{LC}} \tag{3.14}$$

and it is an *all-pass* system (these systems will be encountered again in Chapter 7 and in Chapter 8). These are systems characterized by a frequency response $G(j\omega)$ with $|G(j\omega)| = 1\ \forall\omega$, and a flat magnitude Bode diagram.
We now derive the state-space equations and discuss the Kalman decomposition by using MATLAB®.
Taking into account the following state variables

$$\begin{aligned} x_1 &= i_{L1} \\ x_2 &= v_{C2} \\ x_3 &= i_{L2} \\ x_4 &= v_{C1} \end{aligned} \tag{3.15}$$

the state-space representation of the circuit can be derived by applying Kirchhoff's circuit laws. The following equations are obtained:

$$\begin{aligned} \dot{x}_1 &= -\frac{x_4}{L1} + \frac{u}{L_1} \\ \dot{x}_2 &= -\frac{x_2}{C_2 R} + \frac{x_3}{C_2} - \frac{x_4}{C_2 R} + \frac{u}{R} \\ \dot{x}_3 &= -\frac{x_2}{L2} + \frac{u}{L_2} \\ \dot{x}_4 &= \frac{x_1}{C_1} - \frac{x_2}{C_1 R} - \frac{x_4}{C_1 R} + \frac{u}{R} \end{aligned} \tag{3.16}$$

and

$$y = x_2 + x_4 - u \tag{3.17}$$

From equations (3.16) and (3.17), the state-space matrices are derived:

$$\mathrm{A} = \begin{bmatrix} 0 & 0 & 0 & -\frac{1}{L_1} \\ 0 & -\frac{1}{C_2 R} & \frac{1}{C_2} & -\frac{1}{C_2 R} \\ 0 & -\frac{1}{L_2} & 0 & 0 \\ \frac{1}{C_1} & -\frac{1}{C_1 R} & 0 & -\frac{1}{C_1 R} \end{bmatrix}; \mathrm{B} = \begin{bmatrix} \frac{1}{L_1} \\ \frac{1}{C_2 R} \\ \frac{1}{L_2} \\ \frac{1}{C_1 R} \end{bmatrix}; \mathrm{C} = \begin{bmatrix} 0 & 1 & 0 & 1 \end{bmatrix}; \mathrm{D} = -1 \tag{3.18}$$

Let us now consider $C = 1$, $L = 1$ and therefore $R = 1$ (all expressed in dimensionless units), so that we have

$$\mathrm{A} = \begin{bmatrix} 0 & 0 & 0 & -1 \\ 0 & -1 & 1 & -1 \\ 0 & -1 & 0 & 0 \\ 1 & -1 & 0 & -1 \end{bmatrix}; \mathrm{B} = \begin{bmatrix} 1 \\ 1 \\ 1 \\ 1 \end{bmatrix}; \mathrm{C} = \begin{bmatrix} 0 & 1 & 0 & 1 \end{bmatrix}; \mathrm{D} = -1 \tag{3.19}$$

and let us calculate the minimal realization and the Kalman decomposition through MATLAB (it is in fact obvious that equations (3.19) are not a minimal realization of the system with transfer function (3.14)).
First of all, let us define the system:

```
>> A=[0 0 0 -1; 0 -1 1 -1; 0 -1 0 0; 1 -1 0 -1]
>> B=[1 1 1 1]'
>> C=[0 1 0 1]
>> D=-1
>> system=ss(A,B,C,D)
```

Let us now calculate the reachability, the observability and the Hankel matrix:

```
>> Mc=ctrb(A,B)
>> Mo=obsv(A,C)
>> H= Mo*Mc
```

One obtains:

$$\mathrm{M}_c = \begin{bmatrix} 1 & -1 & 1 & -1 \\ 1 & -1 & 1 & -1 \\ 1 & -1 & 1 & -1 \\ 1 & -1 & 1 & -1 \end{bmatrix}; \mathrm{M}_o = \begin{bmatrix} 0 & 1 & 0 & 1 \\ 1 & -2 & 1 & -2 \\ -2 & 3 & -2 & 3 \\ 3 & -4 & 3 & -4 \end{bmatrix}$$

and

$$\mathrm{H} = \begin{bmatrix} 2 & -2 & 2 & -2 \\ -2 & 2 & -2 & 2 \\ 2 & -2 & 2 & -2 \\ -2 & 2 & -2 & 2 \end{bmatrix}$$

The ranks of these matrices are then computed:

```
>> rank(Mc)
>> rank(Mo)
>> rank(H)
```

M_c has rank one, M_o two, and H one. It can be concluded that the minimal form of the system is first order and that the unobservable part is second order and the uncontrollable part is third order. The transfer function with the command `tf(system)` or `zpk(system)` is also calculated, obtaining $G(s) = \frac{1-s}{1+s}$, as in equation (3.14).
Let us now calculate the minimal realization with the MATLAB command:

```
>> [msystem,U] = minreal(system)
```

This gives the system in minimal form (first order):

$$\mathrm{A}_B = -1 \; ; \mathrm{B}_B = 2 \; ; \mathrm{C}_B = 1 \; ; \mathrm{D}_B = -1 \tag{3.20}$$

Matrix U is the inverse of the state transformation matrix from the original representation to the Kalman decomposition. To obtain this, we have thus to consider (since U is orthogonal):

```
>> T=U'
```
and then to compute:
```
>> Atilde=T'*A*T
>> Btilde=T'*B
>> Ctilde=C*T
```
One obtains

$$\tilde{A} = \begin{bmatrix} -1 & -2 & 0 & 0 \\ 0 & -1 & 0 & 0 \\ 0 & 0 & 0 & 1 \\ 0 & 0 & -1 & 0 \end{bmatrix}; \tilde{B} = \begin{bmatrix} 2 \\ 0 \\ 0 \\ 0 \end{bmatrix}; \tilde{C} = \begin{bmatrix} 1 & 1 & 0 & 0 \end{bmatrix}; \tilde{D} = -1 \qquad (3.21)$$

It should be noted that the system is divided in three parts. They appear in $\tilde{A}, \tilde{B}, \tilde{C}, \tilde{D}$ in the following order: controllable and observable part (first order); uncontrollable and observable part (first order); uncontrollable and unobservable part (second order). This example proves how a fourth-order non-minimal circuit realization is adopted to obtain a first-order all pass system by only using passive components.

3.5 Remarks on Kalman decomposition

The example in Figure 3.5 is very different to the example in Figure 3.1. In the first case, if we consider different values of the parameters (for example $R_1 = 0.8$, $C_1 = 1.1$ and $L_2 = 1.2$), the rank of controllability matrix M_c and observability matrix M_o change. The same structure with different parameters (which may be due to perturbations) behaves very differently. In one case, the transfer function is second order and in the other fourth order. This is because the system consists of two symmetrical parts when the values of resistors, capacitors and inductors are equal, but cease to be when those values change.

Instead, the circuit in Figure 3.1 is always an uncontrollable system, notwithstanding the parameter values. Perturbations in the values cannot make it controllable.

For this reason, we must also know how much a system is effectively controllable and observable, not only by evaluating controllability and observability matrices ranks, but also their singular values. If there are very small singular values compared to others, parametric uncertainties may change the structural properties of controllability and observability of the system.

In addition to whether a system is controllable and/or observable or not, we should also ask how much a system is controllable and observable.

In the next chapters we will examine systems that while controllable and observable can also be decomposed (and so approximated) in a part that mostly influences the input/output properties of the system and in a less important part.

An interesting exercise may be to construct, for example, an unobservable and uncontrollable system of given order and with assigned properties of the controllable and observable subsystem. It is easy to show that, once found, an infinite number of such same order systems can be generated.

3.6 Exercises

1. Calculate the Kalman decomposition for the system with state-space matrices:

$$A = \begin{bmatrix} -2 & 3 & 0 & 0 & 0 \\ 1 & 0 & 0 & 0 & 0 \\ 1 & -1 & 3 & 0 & 0 \\ -2 & 1 & -1 & -1 & 0 \\ -4 & 1 & 2 & -1 & -2 \end{bmatrix}; B = C^T = \begin{bmatrix} 1 \\ 1 \\ 1 \\ 1 \\ 1 \end{bmatrix}$$

2. Calculate the Kalman decomposition for the system with state-space matrices:

$$A = \begin{bmatrix} -2 & 0 & 0 & 0 & 0 \\ 1 & 0 & 0 & 0 & 0 \\ 1 & -1 & 3 & 0 & 0 \\ -2 & 1 & -1 & -1 & 0 \\ -4 & 1 & 2 & -1 & -2 \end{bmatrix}; B = C^T = \begin{bmatrix} 1 \\ 1 \\ 1 \\ 1 \\ 1 \end{bmatrix}$$

3. Calculate the Kalman decomposition for the system with state-space matrices:

$$A = \begin{bmatrix} -1 & 0 & 0 & 0 \\ 0 & -1 & 0 & 0 \\ 0 & 0 & 2 & 0 \\ 1 & 1 & -3 & 2 \end{bmatrix}; B = \begin{bmatrix} 0 \\ 0 \\ 1 \\ 0 \end{bmatrix}; C = \begin{bmatrix} 0 & 1 & 1 & 0 \end{bmatrix}$$

4. Calculate the Kalman decomposition for the system with state-space matrices:

$$A = \begin{bmatrix} -1 & 0 & 0 & 0 \\ 0 & -2 & 0 & 0 \\ 0 & 0 & 3 & 0 \\ 0 & 0 & 0 & 2 \end{bmatrix}; B = \begin{bmatrix} 1 \\ 0 \\ 1 \\ 0 \end{bmatrix}; C = \begin{bmatrix} 1 & 1 & 1 & 1 \end{bmatrix}$$

5. Calculate the Kalman decomposition for the system with state-space matrices:

$$A = \begin{bmatrix} -1 & 0 & 0 & 0 \\ 0 & -2 & 0 & 0 \\ 0 & 0 & 3 & 0 \\ 0 & 0 & 0 & 2 \end{bmatrix}; B = \begin{bmatrix} 1 \\ 1 \\ 1 \\ 1 \end{bmatrix}; C = \begin{bmatrix} 0 & 1 & 0 & 1 \end{bmatrix}$$

6. Calculate the Kalman decomposition for the system with state-space matrices:

$$A = \begin{bmatrix} 2 & 1 & 0 & 0 \\ 0 & -1 & 0 & 0 \\ 0 & 0 & 1 & 1 \\ 1 & 0 & -3 & 4 \end{bmatrix}; B = \begin{bmatrix} 1 \\ 0 \\ 1 \\ 0 \end{bmatrix}; C = \begin{bmatrix} 1 & 1 & 1 & 1 \end{bmatrix}$$

4

Singular value decomposition

CONTENTS

4.1 Singular values of a matrix

Any matrix can be decomposed into three matrices with special properties. This unique decomposition is called singular value decomposition.

Consider matrix $\mathrm{A} \in \mathbb{R}^{m\times n}$ and for clarity suppose $m \geq n$, then there are always three matrices U, Σ and V^T such that matrix A can be written as

$$\mathrm{A} = \mathrm{U}\Sigma\mathrm{V}^T$$

such that U is a unitary matrix of dimensions $m \times m$ (i.e., $\mathrm{U}^T\mathrm{U} = \mathrm{I}$), V is also a unitary matrix, but with dimensions $n \times n$, and Σ matrix $m \times n$ is defined as:

$$\Sigma = \begin{bmatrix} \bar{\Sigma} \\ \mathbf{0} \end{bmatrix}$$

In this last expression matrix $\mathbf{0}$ is a matrix of $(m - n) \times n$ null elements, while matrix $\bar{\Sigma}$ of dimension $n \times n$ is a diagonal matrix. The elements of the diagonal of $\bar{\Sigma}$ in descending order are the *singular values* of matrix A:

$$\bar{\Sigma} = \begin{bmatrix} \sigma_1 & 0 & \dots & 0 \\ 0 & \sigma_2 & \dots & 0 \\ \vdots & \vdots & & \vdots \\ 0 & 0 & \dots & \sigma_n \end{bmatrix}$$

Calculating the singular values of matrix A is simple. They are the square roots of the eigenvalues of matrix $\mathrm{A}^T\mathrm{A}$. Note that, since the eigenvalues of $\mathrm{A}^T\mathrm{A}$ are always non-negative real values, it's always possible to calculate their square root.

That the eigenvalues of $\mathrm{A}^T\mathrm{A}$ are always real and non-negative derives from the fact that A^TA is a symmetrical positive semi-definite matrix. The symmetry of this matrix is immediately verifiable, in fact

$$(A^TA)^T = A^TA.$$

Moreover, if we consider the associated quadratic form $V(\mathbf{x}) = \mathbf{x}^TA^TA\mathbf{x}$ and apply theorem 5, we deduce that A^TA is positive definite if A has no null eigenvalues, otherwise it is positive semi-definite.

To prove that the eigenvalues of A^TA are the squares of the singular values of matrix A, consider the singular value decomposition of its transpose:

$$A^T = V\Sigma^TU^T$$

Now consider A^TA, which is a symmetrical matrix and therefore diagonalizable:

$$A^TA = V\Sigma^TU^TU\Sigma V^T$$

Since U is a unitary matrix, then:

$$A^TA = V\bar{\Sigma}^2V^T \tag{4.1}$$

Expression (4.1) represents the diagonalization of matrix A^TA. From this we can draw two important conclusions:

- $\sigma_i^2 = \lambda_i$ (where λ_i are the eigenvalues of A^TA);
- V is the matrix of the orthonormal eigenvectors of A^TA.

In the same way if we consider AA^T (which is a symmetrical matrix $m \times m$, with a maximum rank of n, since $m \geq n$), we obtain m eigenvalues, of which at most n are non-zero. These n eigenvalues are the squares of the singular values of matrix A:

$$AA^T = U\begin{bmatrix} \bar{\Sigma}^2 & \mathbf{0} \\ \mathbf{0} & \mathbf{0} \end{bmatrix}U^T \tag{4.2}$$

from which we deduce that U is the matrix of the orthonormal eigenvectors of AA^T. The columns of U and V are called left-singular and right-singular vectors.

Singular value decomposition can also be applied when matrix A is complex. In this case, instead of the transpose matrix, the conjugate transpose has to be considered. Even here, the singular values are always real and non-negative.

An example of a complex matrix is the transfer matrix $G(s) = C(sI - A)^{-1}B$. The restriction of $G(s)$ to $s = j\omega$, $G(j\omega)$, is a complex matrix as ω varies. Later on, the importance of singular values of matrix $G(j\omega)$ will be discussed.

MATLAB® exercise 4.1

Consider $A \in \mathbb{C}^{3\times 3}$:

$$A = \begin{bmatrix} j & 1 & 1+j \\ 1-j & 2 & j \\ 5+j & j & 5 \end{bmatrix}$$

The singular values of matrix A can be calculated in MATLAB® by first calculating its conjugate transpose with the command:

```
>> A'
```

One obtains:

$$A^* = \begin{bmatrix} -j & 1 & 1-j \\ 1+j & 2 & -j \\ 5-j & -j & 5 \end{bmatrix}$$

Then, the matrix A^*A is calculated:

```
>> A'*A
```

One obtains an Hermitian matrix:

$$A^*A = \begin{bmatrix} 29 & 3+6j & 25-5j \\ 3-6j & 6 & 1-2j \\ 25+5j & 1+2j & 28 \end{bmatrix}$$

Finally, the square root of the eigenvalues of A^*A are calculated:

```
>> sqrt(eig(A'*A))
```

One gets: $\sigma_1 = 7.4044$, $\sigma_2 = 2.7191$ and $\sigma_3 = 0.8843$.

The same result can be directly obtained by the command:

```
>> svd(A)
```

The `svd` command will be discussed in more details in the MATLAB exercise 4.2.

4.2 Spectral norm and condition number of a matrix

In this section we define what is meant by the spectral norm of a matrix. The norm of a matrix (similar to vector norms) is defined as a non-negative number with these properties:

- $\|A\| \geq 0$ for any matrix;
- $\|A\| = 0$ if and only if $A = 0$;
- $\|A + B\| \leq \|A\| + \|B\|$ (triangular inequality).

All the matrix norms which also have (in addition to their listed properties) the property that the norm of the product of two matrices is less than or equal to the product of the norms of the matrices, i.e.:

$$\|A \cdot B\| \leq \|A\| \cdot \|B\|$$

are defined as consistent norms.

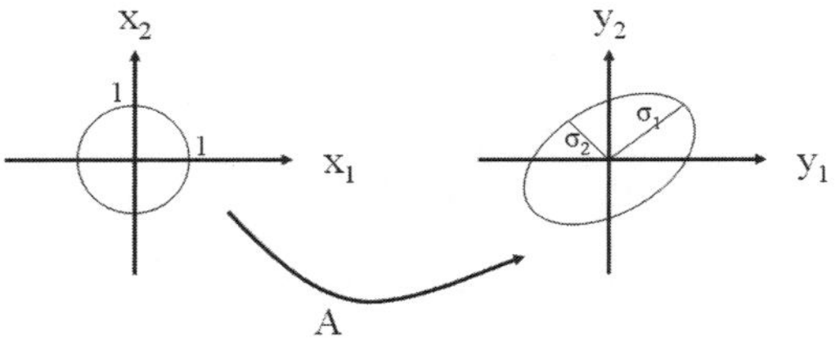

FIGURE 4.1
The unit circumference $\|\mathbf{x}\| = 1$ is mapped into an ellipse in the plane $y_1 - y_2$ by the linear operator A.

Definition 5 (Spectral norm) *The spectral norm of a matrix is the largest singular value of the matrix:*

$$\|\mathrm{A}\|_S = \sigma_1$$

The spectral norm is a consistent norm.

In addition the spectral norm is an induced norm (from the Euclidean vectorial norm) allowing us to clarify its significance. Consider in fact matrix A as a linear operator mapping vector $\mathbf{x} \in \mathbb{R}^n$ into $\mathbf{y} \in \mathbb{R}^n$: $\mathbf{y} = \mathrm{A}\mathbf{x}$. Consider all the vectors $\mathbf{x}$ with unitary norm, i.e., those for which $\|\mathbf{x}\| = 1$ (Euclidean norm, i.e., $\|\mathbf{x}\| = (\sum_{i=1}^{n} x_i^2)^{\frac{1}{2}}$), and calculate the norm of the vectors obtained by the mapping associated with matrix A. The spectral norm corresponds to the maximum of the resulting vector norm:

$$\|\mathrm{A}\|_S = \max_{\|\mathbf{x}\|=1} \|\mathrm{A}\mathbf{x}\|.$$

In the case of matrices $\mathrm{A} \in \mathbb{R}^{2\times 2}$ the spectral norm can be interpreted geometrically. Figure 4.1 shows how a circumference with unitary radius defined by $\|\mathbf{x}\| = 1$ is mapped, using $\mathbf{y} = \mathrm{A}\mathbf{x}$, into an ellipse in the plane $y_1 - y_2$. σ_1 represents the major semi-axis of this ellipse.

Example 4.1

If A is a 2×2 diagonal matrix, e.g., $\mathrm{A} = \begin{bmatrix} \lambda_1 & 0 \\ 0 & \lambda_2 \end{bmatrix}$, then $y_1 = \lambda_1 x_1$ and $y_2 = \lambda_2 x_2$. So:

$$\frac{y_1^2}{\lambda_1^2} = x_1^2$$

and

$$\frac{y_2^2}{\lambda_2^2} = x_2^2$$

Then summing and remembering that $x_1^2 + x_2^2 = 1$ the equations of an ellipse are obtained:

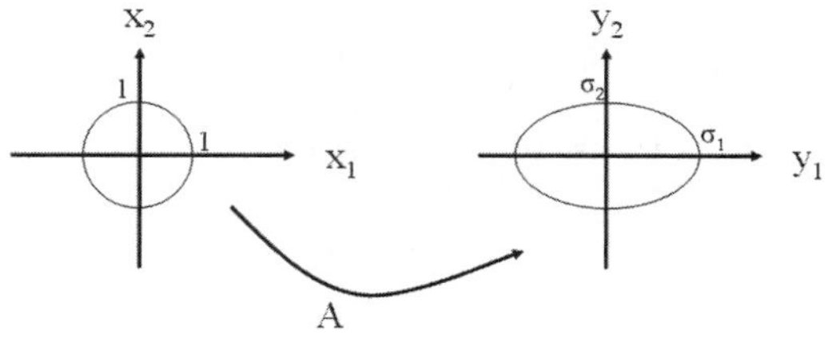

FIGURE 4.2
The unit circumference $\|\mathbf{x}\| = 1$ is mapped into an ellipse in the plane $y_1 - y_2$ by the linear operator A. Example with $\mathrm{A} = \mathrm{diag}(\lambda_1, \lambda_2)$.

$$\frac{y_1^2}{\lambda_1^2} + \frac{y_2^2}{\lambda_2^2} = 1$$

with major semi-axis $\sigma_1 = |\lambda_1|$ and minor semi-axis $\sigma_2 = |\lambda_2|$ as shown in Figure 4.2. Note, the ratio between the maximum and minimum singular value $\frac{\sigma_1}{\sigma_2}$ accounts for the eccentricity of the ellipse. The larger σ_1, the thinner the ellipse. If $\sigma_2 = 0$ (i.e., if matrix A is rank deficient), the ellipse tends to a straight line.

Generally, the invertibility of matrix A depends on the smallest singular value. Since the determinant of a unitary matrix is 1, then

$$\det \mathrm{A} = \det \mathrm{U} \det \Sigma \det \mathrm{V}^T = \det \Sigma = \sigma_1 \cdot \sigma_2 \cdot \ldots \cdot \sigma_n$$

so if the smallest singular value is $\sigma_n \neq 0$, then A is invertible, otherwise it is not.

Moreover, if the matrix rank is $rank(\mathrm{A}) = k$, then k is the number of non-zero singular values.

The ratio between the maximum and minimum singular value of a matrix is therefore a measure of matrix invertibility and is defined as the condition number of a matrix.

Definition 6 (Condition number) *The condition number of an invertible matrix* $\mathrm{A} \in \mathbb{R}^{n \times n}$ *is defined as the ratio between the maximum singular value and the minimum singular value of* A*:*

$$h = \frac{\sigma_1}{\sigma_n}.$$

If h is large, the matrix is ill-conditioned and it is difficult to calculate its inverse. The condition number denotes the invertibility of a matrix, unlike the determinant of a matrix which cannot be considered a measure of matrix invertibility.

Example 4.2

Consider matrices $A = \begin{bmatrix} 1 & 1 \\ 1 & 1 \end{bmatrix}$ and $\tilde{A} = \begin{bmatrix} 1.1 & 1 \\ 0.9 & 1.05 \end{bmatrix}$. A is not invertible since its determinant equals zero, whereas $\tilde{A}$ is invertible ($\det \tilde{A} = 0.255$). However, matrix A can be seen as a perturbation of matrix $\tilde{A}$, i.e.,

$$\tilde{A} + \Delta_A = A$$

with $\Delta_A = \begin{bmatrix} -0.1 & 0 \\ 0.1 & -0.05 \end{bmatrix}$. So, perturbing an invertible matrix ($\tilde{A}$) you obtain a matrix A which is no longer invertible.

Consider instead matrix $\bar{A} = \begin{bmatrix} 10^{-3} & 10^{-3} \\ 5 \cdot 10^{-4} & 10^{-5} \end{bmatrix}$, whose determinant is very small ($\det \bar{A} = -9 \cdot 10^{-8}$), yet the matrix is more robust than $\tilde{A}$ to perturbations that may make it non-invertible. In fact, $h_{\tilde{A}} \simeq 16$ when $h_{\bar{A}} \simeq 4$.

The condition number of a matrix is also related to the issue of the uncertainty of the solution of a system of linear equations. Just consider the system of linear equations

$$A\mathbf{x} = \mathbf{c}$$

with $A \in \mathbb{R}^{n \times n}$, $\mathbf{c} \in \mathbb{R}^n$ known and $\mathbf{x} \in \mathbb{R}^n$ unknown. The solution depends on the invertibility of matrix A. Clearly, any uncertainty about the constant $\mathbf{c}$ will affect the solution. It can be proved that the condition number of matrix A is the link between uncertainty of the known terms and the uncertainty of the solution, i.e.,

$$\frac{||\delta\mathbf{x}||}{||\mathbf{x}||} \leq \frac{\sigma_1}{\sigma_n} \frac{||\delta\mathbf{c}||}{||\mathbf{c}||}$$

where $||\delta\mathbf{x}||$ is the (Euclidean) norm of the error with solution $\mathbf{x}$, caused by uncertainty $\delta\mathbf{c}$ of the known term.

If the condition number of the matrix is large, a little uncertainty in the constant term will cause a large perturbation in the solution.

The condition number of orthonormal matrices has one property easily to be verified.

Theorem 9 *The condition number of a unitary matrix* U *is one.*

In fact, since $U^T U = I$, matrix U has singular values $\sigma_1 = \sigma_2 = \ldots = \sigma_n = 1$ so the condition number is one.

The singular values of a matrix provide upper and lower bounds for the eigenvalues of a matrix. In fact, if σ_1 and σ_n are the maximum and minimum singular values of a matrix A then:

$$\sigma_n \leq \min_i |\lambda_i| \leq \max_i |\lambda_i| \leq \sigma_1 \tag{4.3}$$

To conclude this brief overview of the properties linked to singular values of a matrix, note that, once the singular value decomposition has been calculated, the inverse of a matrix can be immediately computed.

Consider $A = U\Sigma V^T$ and suppose $\sigma_1 \geq \sigma_2 \geq \ldots \geq \sigma_n \neq 0$, the inverse matrix is given by:

$$A^{-1} = V\Sigma^{-1}U^T = V \begin{bmatrix} \frac{1}{\sigma_1} & 0 & \ldots & 0 \\ 0 & \frac{1}{\sigma_2} & \ldots & 0 \\ \vdots & \vdots & & \vdots \\ 0 & 0 & \ldots & \frac{1}{\sigma_n} \end{bmatrix} U^T \tag{4.4}$$

Inversion only requires calculating the inverse of n real numbers.

Finally, singular value decomposition can also be applied to calculating pseudo-inverse matrices. If A is not invertible, at least one singular value is zero: in equation (4.4) only the inverses of non-zero singular values are considered.

MATLAB® exercise 4.2

This exercise explains how to use MATLAB to calculate the singular value decomposition of a matrix.

MATLAB-command for the singular value decomposition is

```
[U,S,V] = svd(A)
```

MATLAB uses an algorithm called the singular value decomposition algorithm which is similar to line elimination for calculating matrix rank. The algorithm is robust and works well also with large matrices. Define matrix A as follows

```
>> A=[0.001 0.001; 0.0001 0.00001]
```

singular values are calculated with command

```
svd(A)
```

You get $\sigma_1 = 0.0014$ and $\sigma_2 = 0.0001$.

Left and right singular vectors A are calculated with command

```
>> [U,S,V]=svd(A)
```

To complete the exercise, verify the following properties:

1. Orthonormality of matrices U and V, using commands:
   ```
   >> U'*U
   >> V'*V
   ```
2. Matrix A is given by $A = U\Sigma V^T$. Use command:
   ```
   U*S*V'
   ```
3. Calculate the inverse of A. Use command
   ```
   >> V*inv(S)*U'
   ```
   and compare the obtained result using command
   ```
   >> inv(A)
   ```

Note. The numerical algorithm for obtaining singular value decomposition was proposed by Golub in 1975.

4.3 Exercises

1. Calculate the singular value decomposition of

$$A = \begin{bmatrix} 0 & 1 & 0 & 0.3 & -3 \\ -1 & -7 & 3 & -7 & -2 \\ 1 & 0.5 & 2 & 1 & 1 \\ 2 & 1 & 0 & 0 & 1 \end{bmatrix}$$

2. Calculate the singular value decomposition of $A = \begin{bmatrix} 2-j & j & 1 \\ -j & 3j & 1 \\ 7 & 1 & 6j \end{bmatrix}$.

3. Calculate the condition number of matrix $A = \begin{bmatrix} -1 & 0.5 & 3 \\ 0.1 & 7 & 1 \\ 3 & -4 & -5 \end{bmatrix}$.

4. Calculate the inverse of $A = \begin{bmatrix} 2 & 0 & 2 & 2 \\ 2 & 5 & -7 & -11 \\ 0 & -2 & 6 & -7 \\ 0 & 2 & -6 & -7 \end{bmatrix}$.

5. Calculate the eigenvalues of

$$A = \begin{bmatrix} 2.5000 & -1.0000 & -3.6416 & -1.6416 \\ -21.9282 & 2.0359 & 13.7439 & 6.2080 \\ 13.4641 & -1.7679 & -8.1927 & -2.9248 \\ -13.4641 & 1.7679 & 11.3343 & 6.0664 \end{bmatrix}$$

and verify that they satisfy equation (4.3).

5

Open-loop balanced realization

CONTENTS

In Chapter 3 we analyzed the Kalman decomposition, which allows us to determine the controllable and observable part of a system. In this Chapter we will deal with determining how controllable and observable the system is.

In simplifying the poles and zeros of a system, Kalman's decomposition examines the exact simplification between pole and zero, whereas the technique in this chapter examines the case in which poles and zeros are very close to each other.

5.1 Controllability and observability gramians

Consider a generic Lyapunov equation

$$\mathrm{A}^T\mathrm{X} + \mathrm{XA} = -\mathrm{Q}$$

if A is the state matrix of an asymptotically stable system, then for theorem 7 there is an integral solution

$$\mathrm{X} = \int_0^\infty e^{\mathrm{A}^T t}\mathrm{Q}e^{\mathrm{A}t}dt.$$

Moreover if Q is positive definite, so too is X.

Now let us consider two particular Lyapunov equations:

$$\mathrm{AX} + \mathrm{XA}^T = -\mathrm{BB}^T \tag{5.1}$$

$$\mathrm{A}^T\mathrm{X} + \mathrm{XA} = -\mathrm{C}^T\mathrm{C} \tag{5.2}$$

associated to the linear time invariant system described by the state matrices (A, B, C). If we assume that the system is asymptotically stable, then each equation has a solution.

In particular, the solution to Lyapunov equation (5.1) is

$$\mathrm{W}_c^2 = \int_0^\infty e^{\mathrm{A}t}\mathrm{B}\mathrm{B}^T e^{\mathrm{A}^T t} dt$$

while the solution to Lyapunov equation (5.2) is

$$\mathrm{W}_o^2 = \int_0^\infty e^{\mathrm{A}^T t}\mathrm{C}^T\mathrm{C} e^{\mathrm{A}t} dt.$$

Note that, generally, for theorem 5 the matrices BB^T and $\mathrm{C}^T\mathrm{C}$ are positive semi-definite matrices. Consider for example a single exit system with $\mathrm{C} = \begin{bmatrix} 1 & 0 \end{bmatrix}$, matrix

$$\mathrm{C}^T\mathrm{C} = \begin{bmatrix} 1 & 0 \\ 0 & 0 \end{bmatrix}$$

has one null and one positive eigenvalue, so it is positive semi-definite. Notwithstanding, it can be proved that if the system is asymptotically stable and controllable, matrix W_c^2 is positive definite. Likewise, if the system is asymptotically stable and observable, matrix W_o^2 is positive definite. These two matrices are called controllability and observability gramians. From a formal point of view, the following are the two gramian definitions.

Definition 7 (Controllability gramian) *Given an asymptotically stable system, the controllability gramian*

$$\mathrm{W}_c^2 = \int_0^\infty e^{\mathrm{A}t}\mathrm{B}\mathrm{B}^T e^{\mathrm{A}^T t} dt$$

is the solution of the following Lyapunov equation:

$$\mathrm{AW}_c^2 + \mathrm{W}_c^2\mathrm{A}^T = -\mathrm{BB}^T \tag{5.3}$$

Definition 8 (Observability gramian) *Given an asymptotically stable system, the observability gramian*

$$\mathrm{W}_o^2 = \int_0^\infty e^{\mathrm{A}^T t}\mathrm{C}^T\mathrm{C} e^{\mathrm{A}t} dt$$

is the solution of the following Lyapunov equation:

$$\mathrm{A}^T\mathrm{W}_o^2 + \mathrm{W}_o^2\mathrm{A} = -\mathrm{C}^T\mathrm{C} \tag{5.4}$$

The notation W_c^2 and W_o^2 (with the square) is to remind us that the matrices are positive definite if the system is asymptotically stable, controllable and observable. In the following example we see the importance of asymptotic stability in obtaining the solution.

Example 5.1

Given a linear first-order system with A = 0, B = 1 and C = 1, the Lyapunov equations (5.3) and (5.4) have no solution. For the system under consideration, in fact, they become $0 \cdot W^2 = 1$ which has no solution. Furthermore, the assumption of asymptotic stability is not verifiable and so this system is only marginally stable. Since there is an eigenvalue at $\lambda = 0$, the necessary conditions for solving a Lyapunov equation of type (5.3) and (5.4) have not been met, that is, there are no eigenvalues on the imaginary axis.

Since the gramians are symmetrical and positive definite, the singular values coincide with the eigenvalues of the matrix. This can be proven by considering a generic positive definite symmetrical matrix W, and recalling that singular values are calculated from the eigenvalues of matrix W^TW. For symmetric matrices $W^TW = W^2$. But since W^2 has λ_i^2 eigenvalues, where λ_i are the eigenvalues of W, it follows that $\sigma_i = \lambda_i$. For this reason, the singular values and eigenvalues of W_c^2 or W_o^2 are indistinguishable.

MATLAB® exercise 5.1

This exercise illustrates the commands for calculating the controllability and observability gramians and whose properties will be discussed.
First example. Consider the linear time-invariant system described by the following state matrices:

$$A = \begin{bmatrix} 0 & 1 & 0 \\ 0 & 0 & 1 \\ -5 & -4 & -3 \end{bmatrix}; B = \begin{bmatrix} 0 \\ 0 \\ 1 \end{bmatrix}; C = \begin{bmatrix} 0 & 1 & 0 \end{bmatrix}$$

Obviously, the system is in canonical control form, so completely controllable, and furthermore asymptotically stable. For this reason the controllability gramian has to be positive definite.
The state matrices are defined by these commands:

```
>> A=[0 1 0; 0 0 1; -5 -4 -3]
>> B=[0; 0; 1]
>> C=[0 1 0]
```

Let's study the observability of the system by calculating the rank of the system's observability matrix, with command:

```
>> rank(obsv(A,C))
```

Since the rank is maximum and the system completely observable, the observability gramian has to be positive definite.
Let's calculate the two gramians by solving the associated Lyapunov equations with the commands:

```
>> Wc2=lyap(A,B*B')
>> Wo2=lyap(A',C'*C)
```

Note, they are two symmetrical matrices. Let's verify that the two gramians are solutions of the associated Lyapunov equations:

```
>> A*Wc2+Wc2*A'+B*B'
>> A'*Wo2+Wo2*A+C'*C
```

In both cases we obtain the expected zero matrix.
Now let's verify that the two gramians are positive definite. To do this we can calculate the eigenvalues of the matrices:

```
>> eig(Wc2)
>> eig(Wo2)
```

and note that they are all positive. The same result can be obtained from Sylvester's proof (a symmetrical matrix is positive definite if all the leading principal minors are positive). For example, for the controllability gramian the test is applied by the commands:

```
>> det(Wc2(1,1))
>> det(Wc2(1:2,1:2))
>> det(Wc2)
```

Gramians can also be calculated with the command **gram**. In this case we have to define an LTI model, for example using the command:

```
>> sistema=ss(A,B,C,0)
```

At this point we can calculate the gramians with the commands:

```
>> Wc2=gram(sistema,'c')
>> Wo2=gram(sistema,'o')
```

Second example. Consider the linear time-invariant system described by the state matrices:

$$A = \begin{bmatrix} -1 & 0 & 0 \\ 1/2 & -1 & 0 \\ 1/2 & 0 & -1 \end{bmatrix} ; B = \begin{bmatrix} 1 & 0 \\ 0 & -1 \\ 0 & 1 \end{bmatrix}$$

In this case the (asymptotically stable) system has two inputs, u_1 and u_2.
Note, the fact that the system is completely controllable is immediately verifiable with command:

```
>> rank(ctrb(A,B))
```

But if we consider only the input u_1 (i.e., $u_2 = 0$) then the system is no more completely controllable. The instruction

```
>> rank(ctrb(A,B(:,1)))
```

gives that the rank is 2. In other words, both inputs are strictly necessary to reach any state of $\mathbb{R}^3$.
Initially, let's suppose both inputs are manipulable and calculate the controllability gramian with command:

```
>> Wc2=lyap(A,B*B')
>> eig(Wc2)
```

We find a positive definite gramian.
Instead, when only u_1 can be used to act on the system ($u_2 = 0$) the gramian

```
>> Wc2=lyap(A,B(:,1)*B(:,1)')
>> eig(Wc2)
```

whose eigenvalues are not all positive (one is zero).

5.2 Principal component analysis

In this paragraph, we will recall the most important properties of principal component analysis.

Consider matrix $W \in \mathbb{R}^{n\times n}$ defined as:

$$W = \int_0^\infty F(t)F(t)^T dt$$

with $F : \mathbb{R} \to \mathbb{R}^{n\times m}$ (time function matrix). The implicit assumption in defining W is that there is an integral.

Consider the singular value decomposition of matrix W. This matrix is certainly positive semi-definite (by definition). So, let's consider a set of orthonormal vectors assigned to the non-negative eigenvalues to obtain singular value decomposition:

$$\mathrm{W} = \mathrm{V}\Sigma\mathrm{V}^T$$

where Σ is the diagonal matrix containing the singular values of W ($\Sigma = \mathtt{diag}\{\sigma_1, \sigma_2, \ldots, \sigma_n\}$) and $\mathbf{v_1}, \mathbf{v_2}, \ldots, \mathbf{v_n}$ constitute the set of orthonormal vectors of W. Since $\mathbf{v_1}, \mathbf{v_2}, \ldots, \mathbf{v_n}$ are orthonormal vectors, they can be used as the orthonormal base of $\mathbb{R}^n$ so that $\mathrm{F}(t)$ represents the sum of n components:

$$\mathrm{F}(t) = \mathbf{v_1}\mathbf{f_1}^T(t) + \mathbf{v_2}\mathbf{f_2}^T(t) + \ldots + \mathbf{v_n}\mathbf{f_n}^T(t)$$

where $\mathbf{f_1}(t), \mathbf{f_2}(t), \ldots, \mathbf{f_n}(t)$ are vectors of m time-dependent elements, and $\mathbf{v_1}, \mathbf{v_2}, \ldots, \mathbf{v_n}$ are vectors of n constant time-independent elements. $\mathbf{f_1}(t), \mathbf{f_2}(t), \ldots, \mathbf{f_n}(t)$ are called principal components of $\mathrm{F}(t)$ and are given by:

$$\mathbf{f_i}^T(t) = \mathbf{v_i}^T\mathrm{F}(t).$$

Principal components have certain properties:

$$\int_0^\infty \mathbf{f_i}^T(t)\mathbf{f_j}(t)dt = 0 \text{ per } i \neq j$$

$$\int_0^\infty \mathbf{f_i}^T(t)\mathbf{f_i}(t)dt = \int_0^\infty \|\mathbf{f_i}\|^2 dt = \sigma_i$$

$$\int_0^\infty \|\mathrm{F}(t)\|_F^2 dt = \sum_{i=1}^n \sigma_i$$

where the Frobenius norm is defined by

$$\|A\|_F = (\sum_{i=1,j=1}^n a_{ij}^2)^{\frac{1}{2}}$$

So, $\mathrm{F}(t)$ can be decomposed into n time function components whose energy can be calculated from the singular values of W.

5.3 Principal component analysis applied to linear systems

Let's apply principal component analysis to linear time-invariant systems which is the same as fixing particular $\mathrm{F}(t)$ with a precise physical meaning in systems theory.

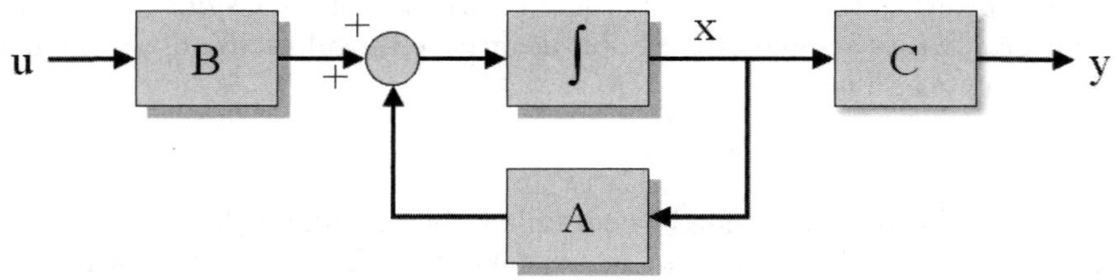

FIGURE 5.1
Linear time-invariant system for defining $\mathrm{F}(t)$ of the controllability gramian.

In particular, given the system in Figure 5.1, imagine applying an input m Dirac impulses and calculating state evolution from null. This would be the same as calculating forced output evolution where $\mathrm{C} = \mathrm{I}$. The state evolution is given by:

$$\mathbf{x}(t) = e^{\mathrm{A}t}\mathrm{B}.$$

By choosing this function as matrix $\mathrm{F}(t)$ in our principal component analysis, where $\mathrm{F}(t) = e^{\mathrm{A}t}\mathrm{B}$, we get the precise definition of the controllability gramian:

$$\mathrm{W}_c^2 = \int_0^\infty \mathrm{F}(t)\mathrm{F}^T(t)dt = \int_0^\infty e^{\mathrm{A}t}\mathrm{B}\mathrm{B}^T e^{\mathrm{A}^T t}dt.$$

In the integral which defines the controllability gramian, matrix $\mathrm{F}(t)$ represents state evolution in response to a Dirac impulse at the input. Controllability subspace X_c is the smallest subspace containing $\Im\mathrm{m}(e^{\mathrm{A}t}\mathrm{B})\ \forall t \in [0, T]$ with $T > 0$.

In the same way, the physical meaning of $\mathrm{F}(t)$ which is in the observability gramian can be calculated. Examine the system in Figure 5.2 with $\mathbf{u} = 0$, apply a vector of n Dirac impulses at the summing node and consider the system response (at null). This is the same as considering the impulse response of a system with $\mathrm{B} = \mathrm{I}$. System response is given by:

$$\mathbf{y}(t) = \mathrm{C}e^{\mathrm{A}t}$$

Note that $\mathrm{F}^T(t) = \mathrm{C}e^{\mathrm{A}t}$ is exactly the term which appears in the observability gramian. In fact the response $\mathbf{y}(t)$ plays a fundamental role in the analysis of the observability properties of the system.

The principal component analysis of function $\mathrm{F}(t) = e^{\mathrm{A}t}\mathrm{B}$ in the controllability gramian W_c^2 and of function $\mathrm{F}^T(t) = \mathrm{C}e^{\mathrm{A}t}$ of the observability gramian W_o^2 allows us to understand, based on the amount of the singular values σ_i relative to W_c^2 or to W_o^2, which components are associated to greater or lesser energy and therefore, those terms which need more energy to be controlled and those that need more energy to be observed.

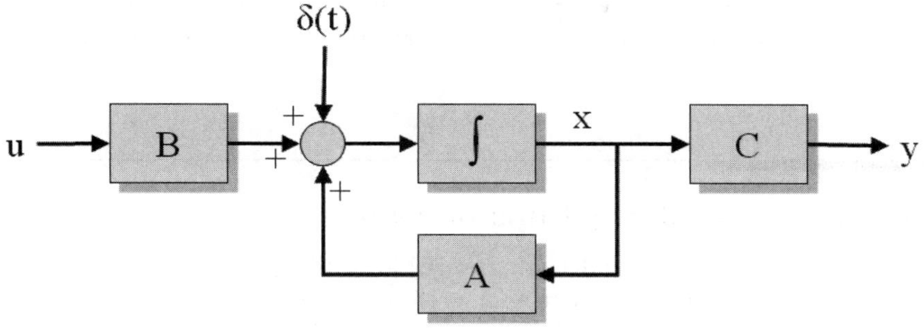

FIGURE 5.2
Linear time-invariant system for defining $\mathrm{F}(t)$ of the observability gramian.

In particular, given the controllability analysis, that function $\mathrm{F}(t) = e^{\mathrm{A}t}\mathrm{B}$, this result is significant. Still considering a controllable system, if $\bar{\mathbf{x}}$ is the state obtained by applying input $\bar{\mathbf{u}}$, the required energy is given by $\int_0^\infty \|\bar{\mathbf{u}}\|^2 dt$. If instead we consider state $\mathbf{x} = \bar{\mathbf{x}} + \Delta\mathbf{x}$, the necessary input would be $\mathbf{u} = \bar{\mathbf{u}} + \Delta\mathbf{u}$. The relation between the energy difference in the two cases and the associated energy at the input $\bar{\mathbf{u}}$ is directly proportional to the relation between $\|\Delta\mathbf{x}\|$ and $\|\mathbf{x}\|$ and the proportionality constant is linked to the relation between the maximum singular value (denoted by σ_{c1}) and the minimum singular value (σ_{cn}) of matrix W_c^2. So, the following expression is valid:

$$\frac{\int_0^\infty \|\bar{\mathbf{u}} - \mathbf{u}\|^2 dt}{\int_0^\infty \|\bar{\mathbf{u}}\|^2 dt} \propto \sqrt{\frac{\sigma_{1c}}{\sigma_{nc}}} \frac{\|\Delta\mathbf{x}\|}{\|\mathbf{x}\|} \tag{5.5}$$

where $\|\cdot\|$ is the Euclidean norm. If the relation $\frac{\sigma_{1c}}{\sigma_{nc}}$ is large, a lot of energy is required to change the state even slightly. The value of the relation $\frac{\sigma_{1c}}{\sigma_{nc}}$ gives an idea of the degree of controllability of the system.

Suppose now the system is observable, the result is analogous:

$$\frac{\int_0^\infty \|\mathbf{y} - \mathbf{y}^*\|^2 dt}{\int_0^\infty \|\mathbf{y}\|^2 dt} \propto \sqrt{\frac{\sigma_{no}}{\sigma_{1o}}} \frac{\|\Delta\mathbf{x}\|}{\|\mathbf{x}\|} \tag{5.6}$$

where $\|\cdot\|$ is the Euclidean norm and $\sigma_{1o}, \ldots, \sigma_{no}$ represent the singular values of matrix $\mathrm{W}_o^2 = \int_0^\infty e^{\mathrm{A}^T t}\mathrm{C}^T\mathrm{C}e^{\mathrm{A}t} dt$.

Note that in this case the proportionality constant is given by the relation between the minimum singular value and the maximum singular value of matrix W_o^2. If the value of this relation is small, i.e., if we have at least one singular gramian value W_o^2 which is smaller than the others, so reconstructing the state from the output terms requires a lot of energy. In fact, ideally it would be good to be able to discriminate a small difference $\|\Delta\mathbf{x}\|$ in the initial condition, so it would be desirable that a small variation significantly affects the first member term. If the $\frac{\sigma_{no}}{\sigma_{1o}}$ ratio is small, a small difference in the initial condition is attenuated. In this case, the variables are hardly observable.

5.4 State transformations of gramians

In this paragraph let's consider any invertible state transformation $\tilde{\mathbf{x}} = \mathrm{T}^{-1}\mathbf{x}$ and see how the gramians change as the reference system changes. Recall that, given system (A, B, C) and applying state transformation $\tilde{\mathbf{x}} = \mathrm{T}^{-1}\mathbf{x}$ we obtain an equivalent system $(\tilde{\mathrm{A}}, \tilde{\mathrm{B}}, \tilde{\mathrm{C}})$ with

$$\begin{aligned} \tilde{\mathrm{A}} &= \mathrm{T}^{-1}\mathrm{AT} \\ \tilde{\mathrm{B}} &= \mathrm{T}^{-1}\mathrm{B} \\ \tilde{\mathrm{C}} &= \mathrm{CT} \end{aligned} \tag{5.7}$$

In the original reference system the gramians are solutions of the Lyapunov equations:

$$\begin{aligned} \mathrm{AW}_c^2 + \mathrm{W}_c^2\mathrm{A}^T &= -\mathrm{BB}^T \\ \mathrm{A}^T\mathrm{W}_o^2 + \mathrm{W}_o^2\mathrm{A} &= -\mathrm{C}^T\mathrm{C} \end{aligned} \tag{5.8}$$

In the new reference system $\tilde{\mathbf{x}}$ the gramians $\tilde{\mathrm{W}}_c^2$ and $\tilde{\mathrm{W}}_o^2$ are solutions of equations:

$$\begin{aligned} \tilde{\mathrm{A}}\tilde{\mathrm{W}}_c^2 + \tilde{\mathrm{W}}_c^2\tilde{\mathrm{A}}^T &= -\tilde{\mathrm{B}}\tilde{\mathrm{B}}^T \\ \tilde{\mathrm{A}}^T\tilde{\mathrm{W}}_o^2 + \tilde{\mathrm{W}}_o^2\tilde{\mathrm{A}} &= -\tilde{\mathrm{C}}^T\tilde{\mathrm{C}} \end{aligned} \tag{5.9}$$

To find the relation between $\tilde{\mathrm{W}}_c^2$ with W_c^2 and $\tilde{\mathrm{W}}_o^2$ with W_o^2, let's apply relations (5.7) to equations (5.9). First, let's consider the controllability gramian:

$$\mathrm{T}^{-1}\mathrm{AT}\tilde{\mathrm{W}}_c^2 + \tilde{\mathrm{W}}_c^2\mathrm{T}^T\mathrm{A}^T(\mathrm{T}^T)^{-1} = -\mathrm{T}^{-1}\mathrm{BB}^T(\mathrm{T}^T)^{-1}$$

Multiplying to the right both members by matrix T^T both, we get:

$$\mathrm{T}^{-1}\mathrm{AT}\tilde{\mathrm{W}}_c^2\mathrm{T}^T + \tilde{\mathrm{W}}_c^2\mathrm{T}^T\mathrm{A}^T = -\mathrm{T}^{-1}\mathrm{BB}^T$$

Multiplying left with matrix T, we get:

$$\mathrm{AT}\tilde{\mathrm{W}}_c^2\mathrm{T}^T + \mathrm{T}\tilde{\mathrm{W}}_c^2\mathrm{T}^T\mathrm{A}^T = -\mathrm{BB}^T \tag{5.10}$$

The expression obtained is equal to the first of the two Lyapunov equations (5.8) if

$$\mathrm{W}_c^2 = \mathrm{T}\tilde{\mathrm{W}}_c^2\mathrm{T}^T$$

from which we get:

$$\tilde{W}_c^2 = T^{-1}W_c^2(T^T)^{-1}$$

Note that $\tilde{W}_c^2$ and W_c^2, are both symmetrical and positive semi-definite (or definite) matrices unlinked by any similarity relation. So, the two gramians do not have the same eigenvalues (nor the same singular values). Therefore, we have reached a very important conclusion: the singular values of the controllability gramian depend on the reference system.

The example in $\mathbb{R}^2$ may be helpful in facilitating an intuitive vision of the problem. In presence of a high condition number, we have a very distorted ellipsoid which reflects an interior unbalance of the system. In this case exist more or less controllable variables. Note that this depends however on the reference system.

With regard to the transformation of the observability gramian, the same reasoning for the controllability gramian can be repeated. Consider the second Lyapunov equation (5.9)(5.9), replacing relations (5.7)(5.7), we obtain:

$$T^TA^T(T^T)^{-1}\tilde{W}_o^2 + \tilde{W}_o^2T^{-1}AT = -T^TC^TCT$$

Multiplying all the terms in this case on the right by T^{-1} and left for $(T^T)^{-1}$, we obtain:

$$A^T(T^T)^{-1}\tilde{W}_o^2T^{-1} + (T^T)^{-1}\tilde{W}_o^2T^{-1}A = -C^TC$$

So, equalling the resulting relation with the second Lyapunov equation (5.8) we obtain:

$$W_o^2 = (T^T)^{-1}\tilde{W}_o^2T^{-1}$$

and therefore

$$\tilde{W}_o^2 = T^TW_o^2T$$

Even with regard to the observability gramian the relation that linking $\tilde{W}_o^2$ with W_o^2 is not a relation of similitude. The observability gramian eigenvalues therefore depend on the reference system.

Summarizing, the relations linking the controllability and observability gramians in two reference systems linked by transformation state $\tilde{\mathbf{x}} = T^{-1}\mathbf{x}$ are:

$$\begin{aligned} \tilde{W}_c^2 &= T^{-1}W_c^2(T^T)^{-1} \\ \tilde{W}_o^2 &= T^TW_o^2T \end{aligned} \tag{5.11}$$

Since the gramian eigenvalues depend on the reference system, also $\frac{\sigma_{1c}}{\sigma_{nc}}$ and $\frac{\sigma_{1o}}{\sigma_{no}}$ depend on the reference system. So, supposing the asymptotically stable system is minimal, the choice of reference system for solving the controllability or observability issue is very significant from the moment that the degree of observability or controllability depends on the reference system. Referring to relations (5.5) (5.5) and (5.6) (5.6), there is a reference system in which all the variables are controllable in the same way. This occurs when the controllability gramian is equal to the identity matrix. In the same way there is a reference system in which all the variables are equally observable, requiring the same energy to reconstruct the initial condition. This occurs when the observability gramian equals the identity matrix. So, controllability and observability (as structural properties) in a system do not change as the reference system changes. What may change is degree of controllability and observability.

At this point we should ask ourselves if there is a reference system in which the two gramians are equal, that is, a reference system in which the degree of controllability and observability are equal. We will see that the answer to this question is positive, and that under the right hypotheses in this reference system it is possible to break this system up into a very controllable and observable part and into a less controllable and observable part.

5.5 Singular values of linear time-invariant systems

We have seen that the controllability and observability gramian eigenvalues depend on the reference system, so they are not system invariants.

As regards the product of the two gramians $\mathrm{W}_c^2\mathrm{W}_o^2$, by applying a transformation state, as the consequence of relations (5.11) (5.11) we find that:

$$\tilde{\mathrm{W}}_c^2\tilde{\mathrm{W}}_o^2 = \mathrm{T}^{-1}\mathrm{W}_c^2\mathrm{W}_o^2\mathrm{T} \tag{5.12}$$

Since $\tilde{\mathrm{W}}_c^2\tilde{\mathrm{W}}_o^2$ and $\mathrm{W}_c^2\mathrm{W}_o^2$ are linked by a relation of similitude, they have the same eigenvalues. The same is true if we consider the observability and controllability gramian product:

$$\tilde{\mathrm{W}}_o^2\tilde{\mathrm{W}}_c^2 = \mathrm{T}^T\mathrm{W}_o^2\mathrm{W}_c^2(\mathrm{T}^T)^{-1} \tag{5.13}$$

The eigenvalues of the gramian product do not change as the reference system changes. Furthermore, since $(\mathrm{W}_c^2\mathrm{W}_o^2)^T = \mathrm{W}_o^2\mathrm{W}_c^2$ the two matrices (the product matrix of the controllability and observability gramian and the product matrix of the observability and controllability gramian) one is the transposition of the other and they have the same eigenvalues. Ultimately, these two matrices are defined (or semi-defined) positive being the product of positive defined (or semi-defined) matrices. More precisely, the hypothesis of asymptotic stability of the system implies that the matrices are semi-defined positive; and if the system is also controllable and observable, the matrices are defined positive.

More generally, these two matrices ($W_c^2 W_o^2$ and $W_o^2 W_c^2$) are not symmetrical, the product of two symmetrical matrices generally not being a symmetrical matrix. Effectively, in the case of the gramian product, without considering particular states:

$$(W_c^2 W_o^2)^T = W_o^2 W_c^2 \neq W_c^2 W_o^2$$

Since the gramian product matrix does not have negative eigenvalues their roots may be found. As before, these roots are system invariants which take the name of singular values of the system.

Definition 9 (Singular values of the system) *Given a linear asymptotically stable system, the singular values of the system $\sigma_1 \geq \sigma_2 \geq \ldots \sigma_n$, are called the eigenvalue roots of controllability and observability gramian product. Singular values are system invariants.*

5.6 Computing the open-loop balanced realization

Once all the required mathematical tools are defined, we can deal with determining a system in which the two gramians are equal and so their controllability and observability are measured by the same parameters since the eigenvalues of W_c^2 coincide with those of W_o^2. Given an asymptotically stable system which is controllable and observable (as we will do in the rest of this chapter, unless otherwise specified), it is therefore possible to distinguish a strongly controllable and observable part as well as a poorly controllable and observable part. Once this system is broken down, we will only consider the strongly controllable and observable part of the system as a low order model. So, let's apply principal component analysis to functions $e^{At}B$ and $(Ce^{At})^T$ making sure to omit the $\mathbf{f}_i^T(t)$ terms associated with small (relatively) singular values. Since singular values represent the energy associated with those components $\mathbf{f}_i^T(t)$ the aim is to be able to identify the low-energy components and omit them. When we apply principal component analysis we have to be sure that the various contributions are weighted only by components $\mathbf{f}_i^T(t)$ and do not depend on $\mathbf{v}_i$. This leads to better specifying the characteristic conditions of this system: the gramians must be equal and diagonal from the moment that the state variables associated with the strongly controllable and observable part can be associated with the weakly controllable and observable part. Ultimately, since the gramian product's eigenvalues equal the squares of the system's singular values, what characterizes it is that the two gramians are diagonal matrices containing the system's singular values. Hence, it is called a balanced open-chain system.

Definition 10 (Open-loop balanced realization) *Given a controllable and observable, linear and asymptotically stable system, the open-loop balanced realization is the realization in which the controllability and observability gramian are equal, diagonal and the diagonal contains the singular values of the system* $W_c^2 = W_o^2 = \mathtt{diag}\{\sigma_1, \sigma_2, \dots, \sigma_n\}$. $W_c^2 = W_o^2 = \mathtt{diag}\{\sigma_1, \sigma_2, \dots, \sigma_n\}$.

Now let's see how to construct such a system, or, which matrix T of transition from the initial reference system to the new balanced one is required.

Consider Lyapunov's solution to the controllability gramian

$$AW_c^2 + W_c^2 A^T = -BB^T$$

Once the W_c^2 solution is found (we know it exists because the system is asymptotically stable), let's consider its singular value decomposition:

$$W_c^2 = V_c \Sigma_c^2 V_c^T$$

where for convenience $\sigma_{1c}^2, \sigma_{2c}^2, \dots, \sigma_{nc}^2$ are the singular values of W_c^2. Note also that $U_c = V_c$ given that the matrix is symmetrical. Recall also that if M is a symmetrical matrix, then $M^T M = MM^T = M^2$ and the eigenvectors of $M^T M$ coincide with those of MM^T.

Consider the transformation state defined by matrix $T_1 = V_c \Sigma_c$ (Σ_c the diagonal matrix which contains the roots of the singular values of W_c^2).

Since $T_1^{-1} = \Sigma_c^{-1} V_c^T$, applying transformation (equation (5.11)) we obtain:

$$\tilde{W}_c^2 = \Sigma_c^{-1} V_c^T W_c^2 V_c \Sigma_c^{-1} = \Sigma_c^{-1} \Sigma_c^2 \Sigma_c^{-1} = I$$

With $\tilde{W}_c^2 = I$ we obtain $\tilde{W}_o^2$ whose eigenvalues equal the square of the singular values of the system, since $\tilde{W}_c^2 \tilde{W}_o^2 = \tilde{W}_o^2$ (this intermediate solution is called *input normal form*).

In this state representation, the system's singular values can be calculated using the Lyapunov equation:

$$\tilde{A}^T \tilde{W}_o^2 + \tilde{W}_o^2 \tilde{A} = -\tilde{C}^T \tilde{C}$$

At this point, considering the singular value break down of the observability gramian, we have:

$$\tilde{W}_o^2 = \tilde{V}_o \Sigma^2 \tilde{V}_o^T$$

where Σ is exactly the diagonal matrix which contains the singular values of system $\sigma_1, \sigma_2, \dots, \sigma_n$.

Consider a second transformation state $T_2 = \tilde{V}_o \Sigma^{-\frac{1}{2}}$. Since $T_2^T = \Sigma^{-\frac{1}{2}} \tilde{V}_o^T$, in the new reference system (let $\bar{\mathbf{x}}$), we have:

$$\bar{W}_o^2 = \Sigma^{-\frac{1}{2}} \tilde{V}_o^T \tilde{V}_o \Sigma^2 \tilde{V}_o^T \tilde{V}_o \Sigma^{-\frac{1}{2}} = \Sigma$$

and

$$\bar{W}_c^2 = \Sigma$$

To balance these out two different transformation states were applied. The overall transformation state is given by $\bar{\mathbf{x}} = T^{-1}\mathbf{x}$. Since $\bar{\mathbf{x}} = T_2^{-1}\tilde{\mathbf{x}} = T_2^{-1}T_1^{-1}\mathbf{x}$, then $T = T_1T_2$.

So, applying transformation state $T = T_1T_2$, a reference system can be obtained in which:

$$\begin{array}{l} \bar{W}_c^2 = \bar{W}_o^2 = \Sigma = \mathtt{diag}\{\sigma_1, \sigma_2, \ldots, \sigma_n\} \\ \bar{W}_c^2\bar{W}_o^2 = \Sigma^2 \end{array} \tag{5.14}$$

MATLAB® exercise 5.2

Now let's find the balanced open-chain form of an asymptotically stable controllable and observable system. Here is an example of the procedure for the following system in state-space form:

$$A = \begin{bmatrix} -1 & 0 & 0 \\ 1/2 & -1 & 0 \\ 1/2 & 0 & -1 \end{bmatrix}; B = \begin{bmatrix} 1 & 0 \\ 0 & -1 \\ 0 & 1 \end{bmatrix}; C = \begin{bmatrix} 0 & 0 & 1 \\ 1 & 1 & 0 \end{bmatrix}$$

Let's follow these steps:

1. Define the system:
```
>> A=[-1 0 0; 1/2 -1 0; 1/2 0 -1]
>> B=[1 0 0; 0 -1 1]'
>> C=[0 0 1; 1 1 0]
>> system=ss(A,B,C,0)
```
2. Verify the hypotheses so the balanced open-chain form (asymptotic stability, controllability, observability) can be calculated:
```
>> eig(A)
>> rank(ctrb(system))
>> rank(obsv(system))
```
3. Calculate the state transformation matrix P_1
```
>> Wc2=lyap(A,B*B')
```
(alternatively `Wc2=gram(system,'c')`)
```
>> [Uc,Sc2,Vc]=svd(Wc2)
>> P1=Vc*sqrt(Sc2)
```
4. Calculate the state-space representation $(\tilde{A}, \tilde{B}, \tilde{C})$ (input normal form)
```
>> Atilde=inv(P1)*A*P1
>> Btilde=inv(P1)*B
>> Ctilde=C*P1
```
To test it we can calculate $\tilde{W}_c^2$ and see if it equals the identity matrix
```
>> Wtildec2=lyap(Atilde,Btilde*Btilde')
```
5. Calculate the transformation matrix T_2
```
>> Wtildeo2=lyap(Atilde',Ctilde'*Ctilde)
>> [Uo,So2,Vo]=svd(Wtildeo2)
>> T2=Vo*(So2)^(-1/4)
```

6. Calculate the open-loop balanced realization

```
>> Abal=inv(T2)*Atilde*T2
>> Bbal=inv(T2)*Btilde
>> Cbal=Ctilde*T2
>>systembal=ss(Abal,Bbal,Cbal,0)
```

To test it, note that the gramians are diagonal and equal to the singular values of the system:

```
>> Wbalc2=gram(systembal,'c')
>> Wbalo2=gram(systembal,'o')
```

The singular values of the system can be found using these instructions:

```
>> Wc2=lyap(A,B*B')
>> Wo2=lyap(A',C'*C)
>> sqrt(eig(Wc2*Wo2))
```

Note that, instead of using this procedure, shown mainly for educational purposes, the balanced form of the system can be found with the instruction:

```
>> [systemb,G,T,Ti]=balreal(system)
```

Note (typing `help balreal`) how in this case the transformation of command `balreal` is $\mathbf{x}_{bil} = \mathrm{T}\mathbf{x}$, i.e., the inverse of the one found in the discussed procedure.

Example 5.2

Consider the linear time-invariant system:

$$\mathrm{A} = \begin{bmatrix} -2 & 4\alpha \\ -\frac{4}{\alpha} & -1 \end{bmatrix}; \mathrm{B} = \begin{bmatrix} 2\alpha \\ 1 \end{bmatrix}; \mathrm{C} = \begin{bmatrix} \frac{2}{\alpha} & -1 \end{bmatrix}$$

with $\alpha \neq 0$. By calculating the transfer function of this system ($G(s) = \frac{3s+18}{s^2+3s+18}$) we can see that the system is controllable, observable and asymptotically stable for any value of α (the transfer function does not depend on α). So it makes sense to calculate the balanced form of this system.
Consider, for example, $\alpha = 10$ and calculate the balanced form according to the procedure in exercise 5.2. For $\alpha = 10$ the system becomes:

$$\mathrm{A} = \begin{bmatrix} -2 & 40 \\ -0.4 & -1 \end{bmatrix}; \mathrm{B} = \begin{bmatrix} 20 \\ 1 \end{bmatrix}; \mathrm{C} = \begin{bmatrix} 0.2 & -1 \end{bmatrix}$$

The controllability and observability gramians (calculated by MATLAB command `gram`) are:

$$\mathrm{W}_c^2 = \begin{bmatrix} 100 & 0 \\ 0 & 0.5 \end{bmatrix}; \mathrm{W}_o^2 = \begin{bmatrix} 0.01 & 0 \\ 0 & 0.5 \end{bmatrix}$$

The singular values of the system are the roots of the eigenvalues of the product between the two gramians and can be calculated through the command `sqrt(eig(Wc2*Wo2))`. So, we obtain $\sigma_1 = 1$ e $\sigma_2 = 0.5$.
Note that in this case the two gramians are already diagonal matrices. Since they are unequal the system unbalanced. Applying the balancing procedure, since the controllability gramian is already diagonal, we obtain $\mathrm{V}_c = \mathrm{I}$ and so:

$$\mathrm{T}_1 = \mathrm{I}\Sigma_c = \sqrt{\mathrm{W}_c^2} = \begin{bmatrix} 10 & 0 \\ 0 & \sqrt{0.5} \end{bmatrix}$$

At this point consider system $(\tilde{\mathrm{A}}, \tilde{\mathrm{B}}, \tilde{\mathrm{C}})$. In this reference system $\tilde{\mathrm{W}}_c^2 = \mathrm{I}$, while from $\tilde{\mathrm{W}}_o^2$ calculations it is equal to:

$$\tilde{\mathrm{W}}_o^2 = \begin{bmatrix} 1 & 0 \\ 0 & 0.25 \end{bmatrix}$$

Matrix T_2 is given by:

$$T_2 = V_o \Sigma_o^{-\frac{1}{2}} = I\Sigma_o^{-\frac{1}{2}} = (\tilde{W}_o^2)^{-1/4} = \begin{bmatrix} 1 & 0 \\ 0 & \frac{1}{\sqrt{0.5}} \end{bmatrix}$$

which yields that $T = T_1 T_2 = \begin{bmatrix} 10 & 0 \\ 0 & 1 \end{bmatrix}$. The particular form of the transformation matrix highlights the fact that in this case it is sufficient to change scale to obtain the open-loop balanced form. Applying transformation $\bar{\mathbf{x}} = T^{-1}\mathbf{x}$ we obtain the balanced form:

$$\bar{A} = \begin{bmatrix} -2 & 4 \\ -4 & -1 \end{bmatrix}; \bar{B} = \begin{bmatrix} 2 \\ 1 \end{bmatrix}; \bar{C} = \begin{bmatrix} 2 & -1 \end{bmatrix}$$

In the most trivial case, the balanced form is obtained by appropriately scaling the state variables (in this case, with $\bar{x}_1 = \frac{1}{10}x_1$ and $\bar{x}_2 = x_2$). To generally determine the balanced form, a reference system change is usually needed. In this case, with diagonal gramians it is sufficient to scale the state variables.
Let's return to the more general case with $\alpha \neq 0$. In this case the gramians are:

$$W_c^2 = \begin{bmatrix} \alpha^2 & 0 \\ 0 & 0.5 \end{bmatrix}; W_o^2 = \begin{bmatrix} \frac{1}{\alpha^2} & 0 \\ 0 & 0.5 \end{bmatrix}$$

Consider the case in which α is very small. The variable's degree of controllability which corresponds to the singular value of W_c^2 equalling $\sigma_2 = \alpha$ is very small. But the larger the variables degree of observability, the smaller its controllability.
This example confirms the importance of having balanced controllability and observability which also helps understand at the same time which variables are less controllable and observable.
In this system the parameter α represents the imbalance between observability and controllability but despite its value the balanced form is equal to:

$$\bar{A} = \begin{bmatrix} -2 & 4 \\ -4 & -1 \end{bmatrix}; \bar{B} = \begin{bmatrix} 2 \\ 1 \end{bmatrix}; \bar{C} = \begin{bmatrix} 2 & -1 \end{bmatrix}$$

So we have a form independent of α which obviously coincides with the case in which $\alpha = 1$, that is when the system variables are observable and controllable in the same way.

5.7 Balanced realization for discrete-time linear systems

Everything we have said until now about linear time-continuous systems can be easily extended to linear time-discrete systems:

$$\begin{cases} \mathbf{x}(k+1) = \tilde{A}\mathbf{x}(k) + \tilde{B}\mathbf{u}(k) \\ \mathbf{y}(k) = \tilde{C}\mathbf{x}(k) \end{cases} \tag{5.15}$$

The Lyapunov equations of the gramians take a slightly different form:

$$\tilde{A}\tilde{W}_c^2\tilde{A}^T - \tilde{W}_c^2 = -\tilde{B}\tilde{B}^T \tag{5.16}$$

$$\tilde{A}^T\tilde{W}_o^2\tilde{A} - \tilde{W}_o^2 = -\tilde{C}^T\tilde{C} \tag{5.17}$$

We are still dealing with linear equations for which if a system is controllable, $\tilde{B}\tilde{B}^T$ is semi-defined positive and $\tilde{W}_c^2$ is defined positive, then the system is asymptotically stable.

Let's hypothesize that the system is asymptotically stable, and $\tilde{W}_c^2$ and $\tilde{W}_o^2$ are dual systems compared to the integral in the case of time continuous systems:

$$\tilde{W}_c^2 = \sum_{i=0}^{\infty} \tilde{A}^i \tilde{B}\tilde{B}^T (\tilde{A}^T)^i \tag{5.18}$$

$$\tilde{W}_o^2 = \sum_{i=0}^{\infty} (\tilde{A}^T)^i \tilde{C}^T \tilde{C}\tilde{A}^i \tag{5.19}$$

The two series converge if the system is asymptotically stable. Defining the balanced form for time-discrete systems is otherwise quite similar to that for time-continuous systems.

There are some algorithms which can solve the Lyapunov equation for time-discrete systems, but normally we use the bilinear transformation $z = \frac{1+s}{1-s}$ to make the time-discrete system a fictitious time-continuous system which is analogous in terms of stability. By applying the bilinear transformation to a time-discrete system in state form ($\tilde{A}$, $\tilde{B}$, $\tilde{C}$, $\tilde{D}$) we obtain a time-continuous equivalent (A, B, C, D) with:

$$\begin{array}{l} A = (I + \tilde{A})^{-1}(\tilde{A} - I) \\ B = \sqrt{2}(I + \tilde{A})^{-1}\tilde{B} \\ C = \sqrt{2}\tilde{C}(I + \tilde{A})^{-1} \\ D = \tilde{D} - \tilde{C}(I + \tilde{A})^{-1}\tilde{B} \end{array} \tag{5.20}$$

The singular values are also invariant according to this transformation.

Once the balanced form of the equivalent time continuous system is calculated, we apply the inverse transformations of (5.20) to obtain the balanced time-discrete system.

MATLAB® exercise 5.3

In this MATLAB® exercise the open-loop balanced realization of a discrete-time linear system is computed. Let the system be described by the following state-space matrices:

$$A = \begin{bmatrix} 0.5 & 0 & 0 \\ 0 & -0.7 & 0 \\ 0 & 0 & 0.3 \end{bmatrix}; B = \begin{bmatrix} 1 \\ 1 \\ 1 \end{bmatrix}; C = \begin{bmatrix} 1 & 1 & 1 \end{bmatrix}; D = 0 \tag{5.21}$$

System (5.21) is stable (having all eigenvalues inside the unitary circle), controllable and observable and therefore the open-loop balanced realization can be calculated.
Let's define the system with the MATLAB commands:

```
>> A=[0.5 0 0; 0 -0.7 0; 0 0 0.3]
>> B=[1; 1; 1]
>> C=[1 1 1]
>> D=0
>> system=ss(A,B,C,D,-1)
```

Notice that, since the sampling time is unspecified, in the MATLAB command $Ts = -1$ has been set. The open-loop balanced realization can be calculated with the `balreal` command as follows:

```
>> [systembal,S]=balreal(system)
```

One obtains:

$$\bar{\mathrm{A}} = \begin{bmatrix} -0.04133 & 0.5468 & -0.02004 \\ 0.5468 & -0.2342 & -0.1298 \\ -0.02004 & -0.1298 & 0.3755 \end{bmatrix}; \quad \bar{\mathrm{B}} = \begin{bmatrix} -1.727 \\ -0.122 \\ -0.04216 \end{bmatrix}; \quad \bar{\mathrm{C}} = \begin{bmatrix} -1.727 & -0.122 & -0.04216 \end{bmatrix}; \quad \bar{\mathrm{D}} = 0 \tag{5.22}$$

5.8 Exercises

1. Given the continuous-time LTI system with state matrices

$$\mathrm{A} = \begin{bmatrix} 0 & 1 \\ -3 & -2 \end{bmatrix}; \mathrm{B} = \begin{bmatrix} 0 \\ 1 \end{bmatrix}; \mathrm{C} = \begin{bmatrix} -1 & 1 \end{bmatrix}$$

calculate the gramians and study system controllability and observability.

2. Calculate analytically the singular values of the system with transfer function $G(s) = \frac{s+3}{(s+1)(s+7)}$.

3. Calculate analytically the singular values of the system with transfer function $G(s) = \frac{s^2+1}{s^2+s+1}$.

4. Calculate the singular values of the system with transfer function $G(s) = -10 + \frac{60s}{s^2+3s+2}$.

5. Calculate the open-loop balanced realization of system with state-space matrices:

$$\mathrm{A} = \begin{bmatrix} -0.5 & -1 & 0 & 0 \\ 1 & -0.5 & 0 & 0 \\ 0 & 0 & -3 & 0 \\ 0 & 0 & 0 & -4 \end{bmatrix}; \mathrm{B} = \begin{bmatrix} 1 \\ -1 \\ -1 \\ 1 \end{bmatrix}; \mathrm{C} = \begin{bmatrix} 0 & 1 & -1 & 1 \end{bmatrix}$$

6

Reduced order models

CONTENTS

In this chapter we will look at constructing a reduced order model from a dynamical system. There are various ways of doing this and below we first calculated a system whose state variables were ordered according to a particular system characteristic. For example, the last chapter showed us how to build a system in which the state variables are ordered by controllability or observability or by both properties (open-loop balanced realization).

The next step is to eliminate the less important state variables from the original system so as to obtain a reduced order model with a lower number of state variables. This can be done by *direct truncation* or by *singular perturbation approximation*. Both techniques will be examined.

Another general aspect of the methods for constructing reduced order models is the requirement to obtain a small error, the error being a certain norm between the original model and the reduced order one (the chosen norm helps classify the approximation methods). The two techniques for constructing a reduced order model will be examined from the point of view of the errors they produce (so at the quality of the model).

This chapter deals with reduced order models with an open-loop balanced form, but other chapters will deal with what other types of reduced order models may be built.

6.1 Reduced order models based on the open-loop balanced realization

The first step in building a reduced order model is to calculate a state-space representation highlighting certain system characteristics. In Chapter 5 we saw that an open-loop balanced form is a state representation in which controllability and observability were measured by the same parameters which highlights that parts of the system are strongly controllable and observable and other parts are weakly controllable and observable.

Let us then consider an asymptotically stable system constructed in a state-space form:

$$\begin{cases} \dot{\mathbf{x}} = \mathrm{A}\mathbf{x} + \mathrm{B}\mathbf{u} \\ \mathbf{y} = \mathrm{C}\mathbf{x} \end{cases} \tag{6.1}$$

such that the controllability and observability gramians, that is the two solutions of Lyapunovs equations

$$\begin{array}{c} \mathrm{AW}_c^2 + \mathrm{W}_c^2\mathrm{A}^T = -\mathrm{BB}^T \\ \mathrm{A}^T\mathrm{W}_o^2 + \mathrm{W}_o^2\mathrm{A} = -\mathrm{C}^T\mathrm{C} \end{array} \tag{6.2}$$

coincide and are diagonal $\mathrm{W}_c^2 = \mathrm{W}_o^2 = \mathrm{diag}\{\sigma_1, \sigma_2, \ldots, \sigma_n\}$.

Let us suppose that

$$\sigma_1 \geq \sigma_2 \geq \ldots \geq \sigma_r \gg \sigma_{r+1} \geq \ldots \geq \sigma_n$$

In this case $\sigma_r \gg \sigma_{r+1}$ highlights a group of strongly controllable and observable variables associated with high singular values compared to those remaining, and a group of weakly controllable and observable variables associated with low singular values. This subdivision of state variables suggests a partition of the system:

$$\begin{array}{l} \begin{bmatrix} \dot{\mathbf{x}}_1 \\ \dot{\mathbf{x}}_2 \end{bmatrix} = \begin{bmatrix} \mathrm{A}_{11} & \mathrm{A}_{12} \\ \mathrm{A}_{21} & \mathrm{A}_{22} \end{bmatrix} \begin{bmatrix} \mathbf{x}_1 \\ \mathbf{x}_2 \end{bmatrix} + \begin{bmatrix} \mathrm{B}_1 \\ \mathrm{B}_2 \end{bmatrix} \mathbf{u} \\ \mathbf{y} = \begin{bmatrix} \mathrm{C}_1 & \mathrm{C}_2 \end{bmatrix} \begin{bmatrix} \mathbf{x}_1 \\ \mathbf{x}_2 \end{bmatrix} \end{array} \tag{6.3}$$

where $\mathbf{x}_1 = \begin{bmatrix} x_1 \\ x_2 \\ \vdots \\ x_r \end{bmatrix}$ are the strongly controllable and observable variables and

$\mathbf{x}_2 = \begin{bmatrix} x_{r+1} \\ x_{r+2} \\ \vdots \\ x_n \end{bmatrix}$ are the weakly controllable and observable variables.

Given this partition, there are two methods to define the reduced order model: by direct truncation or by approximating singular perturbation approximation.

6.1.1 Direct truncation method

In this method, the weakly controllable and observable part is neglected, using the strongly controllable and observable part as the reduced order model. This model has an order of r, and is obtained from equations (6.3) making $\mathbf{x}_2 = 0$:

$$\begin{cases} \dot{\mathbf{x}}_1 = \mathrm{A}_{11}\mathbf{x}_1 + \mathrm{B}_1\mathbf{u} \\ \mathbf{y} = \mathrm{C}_1\mathbf{x}_1 \end{cases} \tag{6.4}$$

Above all, note that the reduced order model is asymptotically stable (the original system is asymptotically stable). If we look at Lyapunov's equation for the subblock $\mathrm{W}_{c1}^2 = \begin{bmatrix} \sigma_1 & 0 & \dots & 0 \\ 0 & \sigma_2 & \dots & 0 \\ \vdots & \vdots & & \vdots \\ 0 & 0 & \dots & \sigma_r \end{bmatrix}$, then

$$\mathrm{A}_{11}\mathrm{W}_{c1}^2 + \mathrm{W}_{c1}^2\mathrm{A}_{11}^T = -\mathrm{B}_1\mathrm{B}_1^T \tag{6.5}$$

From this, we deduce that matrix A_{11} satisfies Lyapunov's equation with $\mathrm{B}_1\mathrm{B}_1^T$ semi-defined positive and solution W_{c1}^2 defined positive. So, according to Lyapunov's second criteria for time-invariant linear systems, the reduced order model is asymptotically stable. For the same reason, even the neglected subsystem is asymptotically stable.

In terms of block diagram (see Figure 6.1), it may be noted that this method neglects the coupling (given by matrices A_{21} and A_{12}) between the weakly controllable and observable parts and the strongly controllable and observable parts, considering the reduced order model in terms only of the strongly controllable and observable part which does not change system stability. So, one property of the direct truncation method is that the reduced order model is asymptotically stable.

It has been said that to evaluate the quality of a reduced order model the error between the original and the reduced order model must be calculated. Let $\mathrm{G}(s) = \mathrm{C}(s\mathrm{I}-\mathrm{A})^{-1}\mathrm{B}$ be the transfer matrix of the original system and $\tilde{\mathrm{G}}(s) = \mathrm{C}_1(s\mathrm{I} - \mathrm{A}_{11})^{-1}\mathrm{B}_1$ be the transfer matrix of the approximated model and let us consider the frequency response of the two models: $\mathrm{G}(j\omega)$ and $\tilde{\mathrm{G}}(j\omega)$. The difference in frequency response can be quantified by the matrix given by the difference between $\mathrm{G}(j\omega)$ and $\tilde{\mathrm{G}}(j\omega)$. Since this matrix is a function of ω, calculating the spectral norm will produce a result which depends on ω. Lets then consider the maximum singular value (i.e., the maximum of the spectral norm) with respect to ω: $\max_\omega \|\mathrm{G}(j\omega) - \tilde{\mathrm{G}}(j\omega)\|_S$. The following result holds:

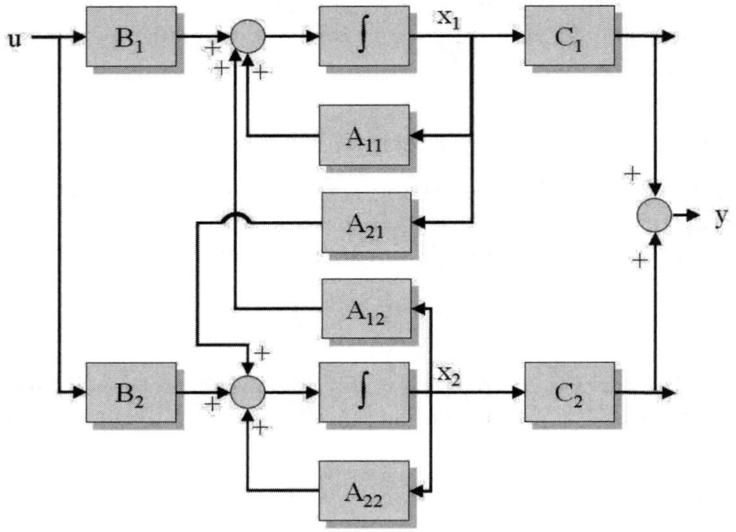

FIGURE 6.1
Block diagram for a partitioned system as in equation (6.3).

$$\max_{\omega} \|\mathrm{G}(j\omega) - \tilde{\mathrm{G}}(j\omega)\|_S \leq 2 \sum_{i=r+1}^{n} \sigma_i \tag{6.6}$$

The norm defined by the first member of equation (6.6) is also known as the H_∞ norm of a transfer matrix.

Definition 11 (H_∞ norm of a system) *The H_∞ norm of a system is defined as the maximum value of the spectral norm of matrix* $\mathrm{G}(j\omega)$ *with respect to* ω*:*

$$\|G(s)\|_\infty = \max_{\omega} \|\mathrm{G}(j\omega)\|_S$$

The error produced by the direct truncation method is less than twice the sum of singular values associated with the variables not considered in the reduced order model. The quality of the model is greater the smaller the singular values $\sigma_{r+1} + \ldots + \sigma_n$ of the neglected part.

6.1.2 Singular perturbation method

In the direct TR method the weakly controllable and observable state variables are totally discarded, their contribution assumed of no relevance. In the singular perturbation method it is supposed that the weakly controllable and observable part is faster than the strongly controllable and observable part (furthermore, we known it is asymptotically stable). What is neglected is the subsystem dynamic, supposing that the variables quickly reach the steady-state. So, imagine that the weakly controllable and observable subsystem evolves so rapidly as to make the state variables $\mathbf{x}_2$ at the steady-state. This is the same as supposing $\dot{\mathbf{x}}_2 = 0$.

Making $\dot{\mathbf{x}}_2 = 0$ in the second equation (6.3), then:

$$\mathbf{x}_2 = \mathrm{A}_{22}^{-1}(-\mathrm{A}_{21}\mathbf{x}_1 - \mathrm{B}_2\mathbf{u}) \tag{6.7}$$

and by substituting in the first equation (6.3) we get

$$\dot{\mathbf{x}}_1 = \mathrm{A}_{11}\mathbf{x}_1 + \mathrm{A}_{12}\mathrm{A}_{22}^{-1}(-\mathrm{A}_{21}\mathbf{x}_1 - \mathrm{B}_2\mathbf{u}) + \mathrm{B}_1\mathbf{u} \tag{6.8}$$

and so

$$\dot{\mathbf{x}}_1 = (\mathrm{A}_{11} - \mathrm{A}_{12}\mathrm{A}_{22}^{-1}\mathrm{A}_{21})\mathbf{x}_1 + (-\mathrm{A}_{12}\mathrm{A}_{22}^{-1}\mathrm{B}_2 + \mathrm{B}_1)\mathbf{u} \tag{6.9}$$

Proceeding in the same way for output $\mathbf{y}$ we get:

$$\mathbf{y} = (\mathrm{C}_1 - \mathrm{C}_2\mathrm{A}_{22}^{-1}\mathrm{A}_{21})\mathbf{x}_1 - \mathrm{C}_2\mathrm{A}_{22}^{-1}\mathrm{B}_2\mathbf{u} \tag{6.10}$$

The reduced order model obtained is thus not strictly proper:

$$\begin{cases} \dot{\mathbf{x}}_1 = \bar{\mathrm{A}}_{11}\mathbf{x}_1 + \bar{\mathrm{B}}_1\mathbf{u} \\ \mathbf{y} = \bar{\mathrm{C}}_1\mathbf{x}_1 + \bar{\mathrm{D}}_1\mathbf{u} \end{cases} \tag{6.11}$$

with

$$\begin{aligned} \bar{\mathrm{A}}_{11} &= \mathrm{A}_{11} - \mathrm{A}_{12}\mathrm{A}_{22}^{-1}\mathrm{A}_{21} \\ \bar{\mathrm{B}}_1 &= -\mathrm{A}_{12}\mathrm{A}_{22}^{-1}\mathrm{B}_2 + \mathrm{B}_1 \\ \bar{\mathrm{C}}_1 &= \mathrm{C}_1 - \mathrm{C}_2\mathrm{A}_{22}^{-1}\mathrm{A}_{21} \\ \bar{\mathrm{D}}_1 &= -\mathrm{C}_2\mathrm{A}_{22}^{-1}\mathrm{B}_2 \end{aligned} \tag{6.12}$$

$(\bar{\mathrm{A}}_{11}, \bar{\mathrm{B}}_1, \bar{\mathrm{C}}_1)$ is an open-loop balanced system as can be verified by calculating the gramians from equations $\bar{\mathrm{A}}_{11}\mathrm{W}_{c1}^2 + \mathrm{W}_{c1}^2\bar{\mathrm{A}}_{11}^T = -\bar{\mathrm{B}}_1\bar{\mathrm{B}}_1^T$ and $\bar{\mathrm{A}}_{11}^T\mathrm{W}_{o1}^2 + \mathrm{W}_{o1}^2\bar{\mathrm{A}}_{11} = -\bar{\mathrm{C}}_1^T\bar{\mathrm{C}}_1$.

So, the system has the same singular values as the strongly controllable and observable part except that it takes into account the asymptotic contribution of the $\mathbf{x}_2$ state variables, which the direct truncation method does not.

Note that the singular perturbation method is a general method and valid for linear and nonlinear systems. Let us consider a generic nonlinear system:

$$\begin{aligned} \dot{\mathbf{x}}_1 &= f_1(\mathbf{x}_1, \mathbf{x}_2) \\ \varepsilon\dot{\mathbf{x}}_2 &= f_2(\mathbf{x}_1, \mathbf{x}_2) \end{aligned} \tag{6.13}$$

If ε is very small, we can suppose that $\varepsilon\dot{\mathbf{x}}_2 \simeq 0$ (i.e., variables evolve much more rapidly than $\mathbf{x}_1$). So $f_2(\mathbf{x}_1, \mathbf{x}_2) = 0$ from which we get $\mathbf{x}_2 = g(\mathbf{x}_1)$ to obtain $\mathbf{x}_1 = f_1(\mathbf{x}_1, g(\mathbf{x}_1))$.

The singular perturbation method can also be applied to discrete-time systems, but the system with state matrix $\bar{\mathrm{A}}_{11}$ is not open-loop balanced.

The error encountered with the singular perturbation method is given by:

$$\max_{\omega} \|\mathrm{G}(j\omega) - \tilde{\mathrm{G}}(j\omega)\|_S \leq 2 \sum_{i=r+1}^{n} \sigma_i \tag{6.14}$$

where $\mathrm{G}(s)$ is the transfer matrix of the original system, whereas $\tilde{\mathrm{G}}(s)$ is the transfer matrix of the reduced order system.

The error expression of the two methods (direct truncation and singular perturbation) are the same. There is, however, a significant difference. It can be demonstrated that the direct truncation method does not preserve static gain whereas the singular perturbation method does. In the former, $\mathrm{G}(j\omega) = \tilde{\mathrm{G}}(j\omega)$ for $\omega \to \infty$, whereas in the latter $\mathrm{G}(0) = \tilde{\mathrm{G}}(0)$. The two methods differ because they are better able to approximate low (the singular perturbation method) or high (the direct truncation method) frequency behavior.

6.2 Reduced order model exercises

MATLAB® exercise 6.1

In this MATLAB® exercise the procedure for obtaining a reduced order model is illustrated through a fourth order system defined by its transfer function:

$$G(s) = \frac{0.5s^4 + 9s^3 + 47.5s^2 + 95s + 62}{(s+1)(s+2)(s+3)(s+4)}$$

The first step is always to define the system:

```
>> n=4
>> s=tf('s')
>>G=(0.5*s^4+9*s^3+47.5*s^2+95*s+62)/((s+1)*(s+2)*(s+3)*(s+4))
```

Then, the balanced form of the system is calculated:

```
>> [system_bal,S]=balreal(G)
```

To verify this, the gramians can be calculated to see if they are diagonal and equal:

```
>> gram(system_bal,'o')
>> gram(system_bal,'c')
```

Now, let us calculate the reduced order model (6.4):

```
>> r=2
>> reducedordersystem=ss(system_bal.a(1:r,1:r),...
   system_bal.b(1:r),system_bal.c(1:r),system_bal.d)
```

and its transfer function:

```
>> tf(reducedordersystem)
```

The approximation error is given by

```
>> DTerror=G-tf(reducedordersystem)
>> directtruncerror=normhinf(G-tf(reducedordersystem))
```

It can be verified that effectively expression (6.6) is valid, by calculating the sum of the discarded singular values:

```
>> error=2*sum(S(r+1:n))
```

Table 6.1 shows the errors by using a reduced order model as r varies.

TABLE 6.1
Errors as order r varies for the reduced order model

r	$\max_\omega \|\|G(j\omega) - \tilde{G}(j\omega)\|\|_S$
1	0.1283
2	0.0037
3	$4.2637 \cdot 10^{-5}$

MATLAB® exercise 6.2

With reference to the system in exercise 6.1, two reduced order models will be constructed using the two methods described and the results will be compared.
Having defined the system as in exercise 6.1, a reduced order model will be constructed where $r = 2$ given that $\sigma_2 = 0.0623 >> \sigma_3 = 0.0018$.
The direct truncation reduced order model is constructed using the following commands:

```
>>reducedordersystem=ss(system_bal.a(1:2,1:2),...
     system_bal.b(1:2),system_bal.c(1:2),system_bal.d)
```

Now, the transfer function of the reduced order model can be calculated as well as the approximation error:

```
>> tf(reducedordersystem)
>> DTerror=G-tf(reducedordersystem)
>> directtruncationerror=normhinf(G-tf(reducedordersystem))
```

Instead, the singular perturbation reduced order model is constructed using the following commands:

```
>> A11=system_bal.a(1:2,1:2);
>> A12=system_bal.a(1:2,3:4);
>> A21=system_bal.a(3:4,1:2);
>> A22=system_bal.a(3:4,3:4);
>> B1=system_bal.b(1:2);
>> B2=system_bal.b(3:4);
>> C1=system_bal.c(1:2);
>> C2=system_bal.c(3:4);
>> D=system_bal.d;
>> A11r=A11-A12*inv(A22)*A21;
>> B1r=-A12*inv(A22)*B2+B1;
>> C1r=C1-C2*inv(A22)*A21;
>> D1r=-C2*inv(A22)*B2+D;
>> SPreducedordersystem=ss(A11r,B1r,C1r,D1r);
```

Analogously, the transfer function of the singular perturbation approximate model can be calculated as well as the error with the following method:

```
>> tf(SPreducedordersystem)
>> SPerror=G-tf(SPreducedordersystem)
>> singularperterror=normhinf(G-tf(SPreducedordersystem))
```

Finally, with command

```
>> ltiview
```

the two reduced order models can be compared. For example, Figure 6.2(a) shows the unit-step response of the original system and the two reduced order models (the three responses are almost indistinguishable).
From the magnitude Bode diagram of the error between the original model and the

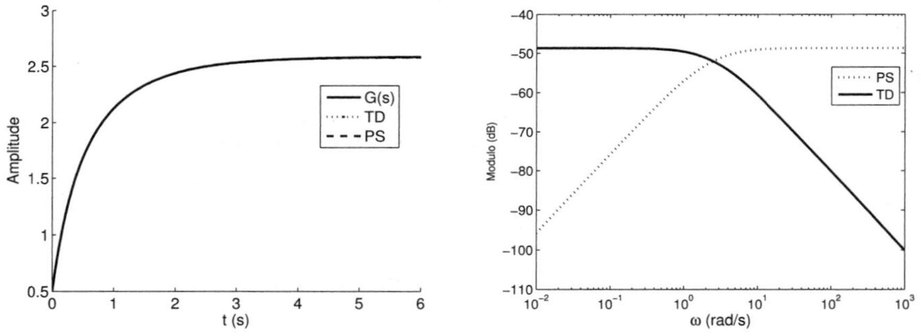

(a) (b)

FIGURE 6.2
(a) Unit-step response of the original system $G(s)$ and of the reduced order models. (b) Magnitude Bode diagram for error between the original model and the reduced order model.

reduced order model, shown in Figure 6.2(b), note that the error obtained in each case was very small, but the approximation of the direct truncation method is better at high frequencies, whereas the singular perturbation method is better at low frequencies.

MATLAB® exercise 6.3

In this further MATLAB® exercise, the use of the `modred` command is described and the properties of the truncated reduced model in the case of continuous-time and discrete-time systems are briefly discussed.
Consider the continuous-time system with state-space matrices:

$$\mathrm{A} = \begin{bmatrix} -1 & 0 & 0 \\ 0 & -2 & 0 \\ 0 & 0 & -3 \end{bmatrix}; \mathrm{B} = \begin{bmatrix} 1 \\ 1 \\ 1 \end{bmatrix}; \mathrm{C} = \begin{bmatrix} 1 & 1 & 1 \end{bmatrix} \tag{6.15}$$

Define the system in MATLAB:

```
>> A=[-1  0 0 ; 0 -2 0; 0 0 -3]
>> B=[1; 1; 1]
>> C=[1 1 1]
>> D=0
>> system=ss(A,B,C,D)
```

and compute the open-loop balanced realization:

```
>> [systembal,S]=balreal(system)
```

Now, let's use the `modred` command. Given a system in balanced form and given a vector indicating which state variables have to be eliminated, the `modred` command gives the reduced order model. To use direct truncation, the option 'Truncate' is used. To use singular perturbation approximation, the option 'MatchDC' (forcing equal DC gains) is used.
Consider a second-order reduced model:

```
>> elim=(S<1e-2)
```

```
>> systemred=modred(systembal,elim,'Truncate')
```
One gets:

$$\bar{A}=\begin{bmatrix} -1.617 & 0.7506 \\ 0.7506 & -2.042 \end{bmatrix};\bar{B}=\begin{bmatrix} -1.682 \\ 0.4087 \end{bmatrix};\bar{C}=\begin{bmatrix} -1.682 & 0.4087 \end{bmatrix};\bar{D}=0 \tag{6.16}$$

Now, let's compute the two gramians
```
>> Wo2=gram(systemred,'o')
>> Wc2=gram(systemred,'c')
```
We get:

$$\bar{W}_c^2=\begin{bmatrix} 0.8751 & 0.0000 \\ 0.0000 & 0.0409 \end{bmatrix};\bar{W}_o^2=\begin{bmatrix} 0.8751 & 0.0000 \\ 0.0000 & 0.0409 \end{bmatrix} \tag{6.17}$$

The singular values of the reduced order system are exactly the same as the original system, as can be observed by comparing the diagonal elements of the gramians with the first two singular values listed in $S=\begin{bmatrix} 0.8751 \\ 0.0409 \\ 0.0006 \end{bmatrix}$.

We now consider a discrete-time example. Let's come back to system (5.21) and apply the `modred` command:
```
>> A=[0.5 0 0; 0 -0.7 0; 0 0 0.3]
>> B=[1; 1; 1]
>> C=[1 1 1]
>> D=0
>> system=ss(A,B,C,D,-1)
>> [systembal,S]=balreal(system)
>> elim=(S<1e-1)
>> systemred=modred(systembal,elim,'Truncate')
>> Wc2=gram(systemred,'c')
>> Wo2=gram(systemred,'o')
```
We get:

$$\bar{W}_c^2=\begin{bmatrix} 3.3064 & 0.0000 \\ 0.0000 & 1.0616 \end{bmatrix};\bar{W}_o^2=\begin{bmatrix} 3.3064 & 0.0000 \\ 0.0000 & 1.0616 \end{bmatrix} \tag{6.18}$$

In this case, the singular values of the reduced order system are not exactly the same as the original system. In fact, the second singular value of the reduced order system is $\sigma_2 = 1.0616$, while the second singular value of the original system (listed in $S=\begin{bmatrix} 3.3065 \\ 1.0621 \\ 0.0244 \end{bmatrix}$) is $\sigma_2 = 1.0621$.

6.3 Exercises

1. Given the system with transfer function

$$G(s) = 1 - \frac{56s^6 + 2676s^4 + 8736s^2 + 1600}{(s+10)^2(s+2)^3(s+1)^2}$$

determine an open-loop balanced realization and an opportune reduced order model.

2. Calculate a reduced order model of the system $G(s) = 200\frac{(s+10)(s^2+s+1)}{(s+5)^3(s+4)(s+2)^2}$.

3. Given the system $G(s) = \frac{s+1}{s^4+4.3s^3+7.92s^2+7.24s+2.64}$ calculate the reduced order models with direct truncation and singular perturbation approximation and compare the models obtained.

4. Given the continuous-time system with state-space matrices:

$$A = \begin{bmatrix} -2 & 1 & 1 & 0 \\ -3 & -5 & 6 & 1 \\ 0 & -1 & -5 & 0 \\ -4 & -5 & -7 & -1 \end{bmatrix}; \quad B = \begin{bmatrix} 0 \\ 1 \\ 0 \\ 1 \end{bmatrix}; \quad C = \begin{bmatrix} 1 & 2 & 1 & 2 \end{bmatrix}$$

   calculate the reduced order models with direct truncation and singular perturbation approximation and the error and compare the rise and settling time of the reduced order models and original system.

5. Given system $G(s) = \frac{s}{s^4+3s^3+4.1s^2+4.1s+0.2}$, choose the order of the reduced model to guarantee an error between nominal model and approximation not larger than 0.02.

7

Symmetrical systems

CONTENTS

There is a particular class of linear time-invariant systems with the advantage of having a symmetrical transfer matrix $G(s)$. They are called symmetrical systems. Obviously the definition makes sense if the transfer matrix is squared, that is, that the number of inputs equals the number of outputs.

Definition 12 (Symmetrical system) *A linear time-invariant system with the number of inputs equalling the number of outputs ($m = p$) is symmetrical if its transfer matrix $G(s)$ is such that $G(s) = G^T(s)$.*

SISO systems are an example of symmetrical systems for which $m = p = 1$ and the transfer matrix is such that $G(s) = G^T(s)$ as it is a scalar function.

7.1 Reduced order models for SISO systems

The issue of reduced order models of symmetrical systems can be faced by reasoning in terms of the energy of impulse response. Given an asymptotically stable symmetrical SISO system

$$\begin{cases} \dot{\mathbf{x}} = \mathrm{A}\mathbf{x} + \mathrm{B}\mathbf{u} \\ \mathbf{y} = \mathrm{C}\mathbf{x} \end{cases} \tag{7.1}$$

and considering impulse response $y(t) = \mathrm{C}e^{\mathrm{A}t}\mathrm{B}$, the energy associated is defined by $E = \int_0^\infty y(t)^T y(t)dt$. This is a finite integral since the system is asymptotically stable and so the impulse response tends to zero.

Furthermore, this energy is linked to the observability gramian:

$$E = \int_0^\infty \mathrm{B}^T e^{\mathrm{A}^T t}\mathrm{C}^T\mathrm{C}e^{\mathrm{A}t}\mathrm{B}dt = \mathrm{B}^T \int_0^\infty e^{\mathrm{A}^T t}\mathrm{C}^T\mathrm{C}e^{\mathrm{A}t}dt\mathrm{B} = \mathrm{B}^T W_o^2 \mathrm{B}$$

Since the impulse response does not vary as the reference system varies, neither does the energy depend on the reference system used. In particular, in the case of an open-loop balanced form $(\bar{\mathrm{A}}, \bar{\mathrm{B}}, \bar{\mathrm{C}})$ the energy expression is:

$$E = \bar{\mathrm{B}}^T \bar{W}_o^2 \bar{\mathrm{B}} = \sum_{i=1}^n \sigma_i \bar{b}_i^2 \tag{7.2}$$

Expression (7.2) suggests that to correctly reduce system order, it is better to consider conditions on the various terms of the impulse response energy $E = \sigma_1 \bar{b}_1^2 + \sigma_2 \bar{b}_2^2 + \ldots + \sigma_n \bar{b}_n^2$ rather than the condition $\sigma_r \gg \sigma_{r+1}$, neglecting terms with relations of the type $\sigma_r \bar{b}_r^2 \gg \sigma_{r+1}\bar{b}_{r+1}^2$, or in other words variables which contribute less energy.

MATLAB® exercise 7.1

Let us consider two SISO systems: $G_1(s) = \frac{\frac{s}{0.9}+1}{(s+1)(\frac{s}{10}+1)}$ and $G_2(s) = \frac{\frac{s}{9}+1}{(s+1)(\frac{s}{10}+1)}$. Before looking at the open-loop balanced reduced order model, let us make some preliminary considerations. The two systems are asymptotically stable and have a pole at $s = -1$ and another at $s = -10$. For both systems, the impulse response is of the type:

$$G(s) = \frac{A}{s+1} + \frac{B}{s+10} \Rightarrow y(t) = Ae^{-t} + Be^{-10t} \tag{7.3}$$

Obtaining a reduced order model requires neglecting one of the two system modes, the choice of which is based not only on neglecting the mode associated with the fastest eigenvalue, but also on the associated residue, the A and B values. Besides, notice that system $G_1(s)$ has a zero in $z = -0.9$ the effect of which is to cancel out nearly all the dynamics at pole $s = -1$, whereas the zero of system $G_2(s)$ is very close to pole $s = -10$ making residue B small. Effectively, $A = -0.1235$ and $B = 11.2346$ for system $G_1(s)$, whereas for $G_2(s)$ $A = 0.9877$ and $B = 0.1235$.

Now, let us consider the open-loop balanced system using MATLAB® commands:

```
>> s=tf('s')
>> G1=(s/0.9+1)/(s+1)/(s/10+1)
>> G2=(s/9+1)/(s+1)/(s/10+1)
>>  [system1b,S1]=balreal(G1)
>>  [system2b,S2]=balreal(G2)
```

and consider the reduced order models obtained by direct truncation

```
>> reducedsystem1=ss(system1b.a(1,1),system1b.b(1),
        system1b.c(1),0)
>> tf(reducedsystem1)
>>reducedsystem2=ss(system2b.a(1,1),system2b.b(1),
        system2b.c(1),0)
>> tf(reducedsystem2)
```

The transfer functions of the approximated models are: $\tilde{G}_1(s) = \frac{11.17}{s+10.29}$ and $\tilde{G}_2(s) =$

$\frac{1.029}{s+1.038}$. Note that in the first case the reduced order model pole is very close to $s=-10$, whereas in the second case the pole is very close to $s=-1$ which agree with earlier considerations. Note also that, notwithstanding the two original systems had the same static (unitary) gain, the reduced order models (obtained by direct truncation) no longer have the same gain.
As regards the system's singular values $\sigma_1=0.5428$, $\sigma_2=0.0428$ in the first case and $\sigma_1=0.4959$, $\sigma_2=0.0041$ in the second, may it be concluded that the approximation is better in the second case? Let us look at the impulse response energy. The various $\sigma_i\bar{b}_i^2$ terms can be obtained from MATLAB® commands:

```
>> S1(:).*(sistema1b.B(:).^2)
>> S2(:).*(sistema2b.B(:).^2)
```

We get $\sigma_1\bar{b}_1^2=6.0636$, $\sigma_2\bar{b}_2^2=0.0026$ for $G_1(s)$ and $\sigma_1\bar{b}_1^2=0.5103$, $\sigma_2\bar{b}_2^2=0.0003$ for $G_2(s)$. It may be concluded that in terms of impulse response energy, the two approximations are equivalent.

7.2 Properties of symmetrical systems

Given that $G(s)=\mathrm{C}(s\mathrm{I}-\mathrm{A})^{-1}\mathrm{B}$ and $G^T(s)=\mathrm{B}^T((s\mathrm{I}-\mathrm{A})^T)^{-1}\mathrm{C}^T$, it can be verified that for a system to be symmetrical it suffices that $\mathrm{B}=\mathrm{C}^T$ and $\mathrm{A}^T=\mathrm{A}$.

More generally, we will see that if a system is symmetrical then matrices B and C and matrices A^T and A are linked by certain relations which involve an invertible and symmetrical matrix, T. To obtain these relations let us consider $G^T(s)$:

$$G^T(s)=\mathrm{B}^T((s\mathrm{I}-\mathrm{A})^T)^{-1}\mathrm{C}^T$$

If a matrix is invertible and symmetrical ($\mathrm{I}=\mathrm{T}\mathrm{T}^{-1}$), then:

$$G^T(s)=\mathrm{B}^T(s\mathrm{T}\mathrm{T}^{-1}-\mathrm{A}^T)^{-1}\mathrm{C}^T=\mathrm{B}^T\mathrm{T}(s\mathrm{I}-\mathrm{T}^{-1}\mathrm{A}^T\mathrm{T})^{-1}\mathrm{T}^{-1}\mathrm{C}^T$$

Equalling this expression with $G(s)=\mathrm{C}(s\mathrm{I}-\mathrm{A})^{-1}\mathrm{B}$ we find that:

$$\begin{aligned}\mathrm{C}&=\mathrm{B}^T\mathrm{T}\\ \mathrm{B}&=\mathrm{T}^{-1}\mathrm{C}^T\\ \mathrm{A}&=\mathrm{T}^{-1}\mathrm{A}^T\mathrm{T}\end{aligned} \tag{7.4}$$

The first two relations (7.4) are equivalent. The transpose of the first relation is:

$$\mathrm{C}^T=\mathrm{T}^T\mathrm{B}\Rightarrow\mathrm{B}=(\mathrm{T}^T)^{-1}\mathrm{C}^T\Rightarrow\mathrm{B}=\mathrm{T}^{-1}\mathrm{C}^T$$

For symmetrical controllable and observable systems there exists a matrix T which links matrices B and C and which can easily be obtained from the observability and controllability matrices. Assuming that the system is symmetrical and in minimal form, in fact one gets:

$$M_o^T = \begin{bmatrix} \mathrm{C} \\ \mathrm{CA} \\ \vdots \\ \mathrm{CA}^{n-1} \end{bmatrix}^T = \begin{bmatrix} \mathrm{C}^T & \mathrm{A}^T\mathrm{C}^T & \dots & (\mathrm{A}^T)^{n-1}\mathrm{C}^T \end{bmatrix} =$$

$$= \begin{bmatrix} \mathrm{TB} & \mathrm{TAB} & \dots & \mathrm{T}(\mathrm{A}^T)^{n-1}\mathrm{B} \end{bmatrix} = \mathrm{TM}_c \Rightarrow \mathrm{T} = \mathrm{M}_o^T\mathrm{M}_c^{-1}$$

MATLAB® exercise 7.2

Consider the continuous-time LTI system:

$$\mathrm{A} = \begin{bmatrix} -4 & -1.5 & -1.5 \\ -5 & -5.5 & -0.5 \\ -1 & 1.5 & -3.5 \end{bmatrix}; \mathrm{B} = \begin{bmatrix} 0.5 & 0.55 \\ 1.5 & -1.35 \\ -1.5 & 0.45 \end{bmatrix};$$
$$\mathrm{C} = \begin{bmatrix} 3 & 0 & -1 \\ 5.4 & -1.8 & -0.8 \end{bmatrix}; \mathrm{D} = \begin{bmatrix} 0 & 0 \\ 0 & 0 \end{bmatrix} \quad (7.5)$$

```
>> A=[-4 -1.5 -1.5; -5 -5.5 -0.5; -1 1.5 -3.5]
>> B=[0.5 0.55;  1.5 -1.35; -1.5 0.45]
>> C=[3 0 -1; 5.4 -1.8 -0.8]
>> D=zeros(2)
>> system=ss(A,B,C,D)
```

By calculating the transfer matrix of the system with the command

```
>> tf(system)
```

one obtains

$$\mathrm{G}(s) = \begin{bmatrix} \frac{3s^2+26s+47}{s^3+13s^2+47s+35} & \frac{1.2s^2+17.2s+64}{s^3+13s^2+47s+35} \\ \frac{1.2s^2+17.2s+64}{s^3+13s^2+47s+35} & \frac{5.04s^2+56.24s+147.2}{s^3+13s^2+47s+35} \end{bmatrix} \quad (7.6)$$

So, since $G(s) = G^T(s)$, the system is symmetrical. Let's now calculate matrix T. We calculate the controllability and the observability matrices and then extract from these two invertible 3×3 blocks taking the first three columns in M_c, or the first three rows in M_o (which are linear independent):

```
>> Mc=ctrb(AA,BB)
>> Mcr=Mc(1:3,1:3)
>> Mo=obsv(AA,CC)
>> Mor=Mo(1:3,1:3)
>> T=Mor'*inv(Mcr)
```

One obtains a symmetrical matrix:

$$\mathrm{T} = \begin{bmatrix} 9.0000 & 0 & 1.0000 \\ 0.0000 & 2.0000 & 2.0000 \\ 1.0000 & 2.0000 & 3.0000 \end{bmatrix} \quad (7.7)$$

We can now verify that equations (7.4) hold:

```
>> B'*T
>> C
>> inv(T)*A'*T
>> A
```

7.3 The cross-gramian matrix

For symmetrical systems, the product of matrices B and C which produces an nxn matrix can be defined. For symmetrical systems, another Lyapunov equation, called cross-gramian equation, can be also defined:

$$\mathrm{AW}_{co} + \mathrm{W}_{co}\mathrm{A} = -\mathrm{BC} \tag{7.8}$$

In the most general case, BC is not a symmetrical matrix, so the solution to Lyapunov equation (7.8) may not be symmetrical. Furthermore, nothing is known about whether it is defined positive or not.

If the system is asymptotically stable, the solution W_{co} to Lyapunov equation (7.8) can be expressed in integral form:

$$\mathrm{W}_{co} = \int_0^\infty e^{\mathrm{A}t}\mathrm{BC}e^{\mathrm{A}t}dt$$

7.4 Relations between W_c^2, W_o^2 and W_{co}

The cross-gramian is linked to the controllability and observability gramians by matrix T. To obtain these relations, let us consider Lyapunov equations for gramians:

$$\begin{aligned} \mathrm{AW}_c^2 + \mathrm{W}_c^2\mathrm{A}^T &= -\mathrm{BB}^T \\ \mathrm{A}^T\mathrm{W}_o^2 + \mathrm{W}_o^2\mathrm{A} &= -\mathrm{C}^T\mathrm{C} \end{aligned}$$

and in the second equation let us substitute (7.4):

$$\mathrm{TAT}^{-1}\mathrm{W}_o^2 + \mathrm{W}_o^2\mathrm{A} = -\mathrm{TBC}$$

Multiplying left by matrix T^{-1} we obtain

$$\mathrm{AT}^{-1}\mathrm{W}_o^2 + \mathrm{T}^{-1}\mathrm{W}_o^2\mathrm{A} = -\mathrm{BC}$$

Comparing the result with the cross-gramian equation (7.8) we obtain:

$$\mathrm{W}_{co} = \mathrm{T}^{-1}\mathrm{W}_o^2$$

Analogously, starting with Lyapunov's equation for the controllability gramian we find:

$$\mathrm{AW}_c^2 + \mathrm{W}_c^2\mathrm{TAT}^{-1} = -\mathrm{BCT}^{-1}$$

$$\mathrm{AW}_c^2\mathrm{T} + \mathrm{W}_c^2\mathrm{TA} = -\mathrm{BC}$$

$$\Rightarrow \mathrm{W}_{co} = \mathrm{W}_c^2 \mathrm{T}$$

At this point notice that:

$$\mathrm{W}_c^2 \mathrm{W}_o^2 = \mathrm{W}_{co}^2$$

This relation produces an important result for the cross-gramian eigenvalues.

Theorem 10 *The eigenvalues of* W_{co} *in the modulus are equal to the system singular values:* $|\lambda_i(\mathrm{W}_{co})| = \sigma_i$.

In fact, relation $\mathrm{W}_c^2 \mathrm{W}_o^2 = \mathrm{W}_{co}^2$ shows us that the eigenvalues of the square of W_{co} equal the square of the system singular values and therefore the eigenvalues of W_{co}, which may be positive or negative, are in the modulus equal to the system singular values, that is $\lambda_i(\mathrm{W}_{co}) = \pm\sigma_i$.

Often this relation is expressed in a matrix form. If $\Sigma = \begin{bmatrix} \sigma_1 & & & \\ & \sigma_2 & & \\ & & \ddots & \\ & & & \sigma_n \end{bmatrix}$ the diagonal matrix formed by the system singular values, and Λ is the matrix formed by the eigenvalues (in decreasing order of their modulus) $\Lambda = \begin{bmatrix} \lambda_1 & & & \\ & \lambda_2 & & \\ & & \ddots & \\ & & & \lambda_n \end{bmatrix}$, then we can write:

$$\Lambda = \mathrm{S}\Sigma$$

where S is an appropriate matrix (also diagonal) the signature matrix whose diagonal components are either $+1$ o -1.

Since the eigenvalues of W_{co} are in the modulus equal to the system singular values, also the eigenvalues of W_{co} are invariants of the systems. The invariance of the W_{co} eigenvalues can be also proved by seeing how the cross-gramian varies as the reference system varies.

Consider the cross-gramian equation in the new reference system defined by state transformation $\tilde{\mathbf{x}} = \bar{\mathrm{T}}^{-1}\mathbf{x}$

$$\tilde{\mathrm{A}}\tilde{\mathrm{W}}_{co} + \tilde{\mathrm{W}}_{co}\tilde{\mathrm{A}} = -\tilde{\mathrm{B}}\tilde{\mathrm{C}}$$

and then apply relation (5.7):

$$\bar{\mathrm{T}}^{-1}\mathrm{A}\bar{\mathrm{T}}\tilde{\mathrm{W}}_{co} + \tilde{\mathrm{W}}_{co}\bar{\mathrm{T}}^{-1}\mathrm{A}\bar{\mathrm{T}} = -\bar{\mathrm{T}}^{-1}\mathrm{B}\mathrm{C}\bar{\mathrm{T}}$$

$$\Rightarrow \mathrm{A}\bar{\mathrm{T}}\tilde{\mathrm{W}}_{co}\bar{\mathrm{T}}^{-1} + \bar{\mathrm{T}}\tilde{\mathrm{W}}_{co}\bar{\mathrm{T}}^{-1}\mathrm{A} = -\mathrm{B}\mathrm{C}$$

from which we obtain:

$$W_{co} = \bar{T}\tilde{W}_{co}\bar{T}^{-1} \tag{7.9}$$

As opposed to what happens with controllability and observability gramians, in this case we find a relation of similitude. So, the W_{co} eigenvalues do not depend on the reference system.

We have seen that for a symmetrical system in whatever state representation that matrices B and C are linked by relations which depend on matrix T.

In the reference system where the system is open-loop balanced this is the signature matrix. Lets now prove this important result. Preliminarily note that $S^{-1} = S$.

Theorem 11 *In a symmetrical open-loop balanced SISO system,* $T = S$ *in (7.4), i.e.:*

$$\begin{aligned} C &= B^T S \\ B &= S C^T \\ A &= S A^T S \end{aligned}$$

Proof 4 *Lyapunov equations for controllability and observability gramians in an open-loop balanced reference system* ($W_c^2 = W_o^2 = \Sigma$) *become:*

$$\begin{aligned} A\Sigma + \Sigma A^T &= -BB^T \\ A^T\Sigma + \Sigma A &= -C^T C \end{aligned}$$

Subtracting the second equation from the first we obtain:

$$(A - A^T)\Sigma + \Sigma(A^T - A) = -(BB^T - C^T C)$$

The terms on the diagonal in the first member are all zero, whereas for those on the second, the i-th diagonal term is $d_{ii} = -b_i^2 + c_i^2$. *From this we find that* $b_i^2 = c_i^2$ *and so* $b_i = \pm c_i$, *a relation which in matrix form can be expressed through the signature matrix* $B = SC^T$. *It therefore follows, given the uniqueness of* T, *that* $T = S$, *the theorem's thesis.*

The case is particularly interesting when the signature matrix equals the identity matrix, $S = I$, or rather when all W_{co} eigenvalues are positive. In this case $B = C^T$ and $A = A^T$. At the beginning of this chapter we saw that this was a sufficient condition (but not necessary) for a system's symmetry.

Generally, matrix S has a certain number of components equal to +1 (indicated by n_+) and a certain number equal to -1 (n_-). Obviously the sum of the two equals the system order. It can be demonstrated that the difference between these two numbers is related to an important index, which will be now introduced.

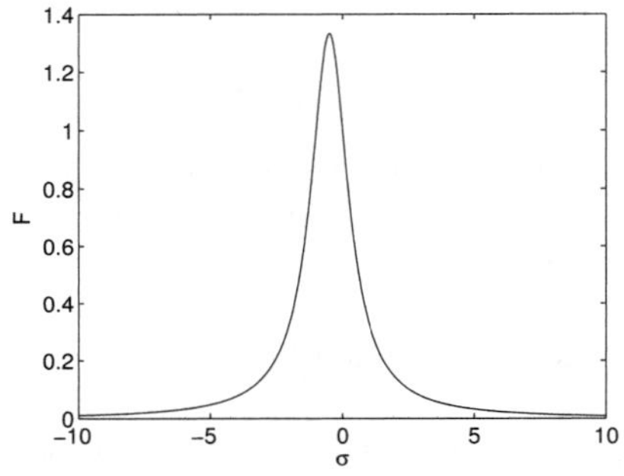

FIGURE 7.1
Plot of $F(\sigma) = \frac{1}{\sigma^2+\sigma+1}$ vs. σ.

Definition 13 (Cauchy index) *The Cauchy index of a real rational function $f(\lambda)$ defined in the interval $[\alpha, \beta]$ is:*

$$I_\alpha^\beta = N_1 - N_2 \tag{7.10}$$

where N_1 is the number of jumps of the function from $-\infty$ to $+\infty$ in the interval $[\alpha, \beta]$ and N_2 is the number of jumps of the function from $+\infty$ to $-\infty$ in the same interval.

The definition of the Cauchy index can be extended to SISO systems with transfer function $F(s)$ when $s = \sigma$, with $\sigma \in \mathbb{R}$. When the interval $(-\infty, \infty)$ is considered, the Cauchy index I_C is given by the number of positive eigenvalues of W_{co} minus the number of negative eigenvalues of W_{co} of a minimal realization of $F(s)$. Also Cauchy index is a system invariant as follows from its definition.

Example 7.1

Consider the system with transfer function $F(s) = \frac{1}{s^2+s+1}$. Consider now $s = \sigma$ with $\sigma \in \mathbb{R}$ and $F(\sigma) = \frac{1}{\sigma^2+\sigma+1}$. The plot of $F(\sigma)$ vs. σ is shown in Figure 7.1, from which it is immediate to derive that $I_{-\infty}^{\infty} = 0$.
Moreover, a minimal realization of $F(s)$ is:

$$A = \begin{bmatrix} 0 & 1 \\ -1 & -1 \end{bmatrix}; B = \begin{bmatrix} 0 \\ 1 \end{bmatrix}; C = \begin{bmatrix} 1 & 0 \end{bmatrix}$$

From this realization we calculate the cross-gramian W_{co}:

$$W_{co} = \begin{bmatrix} 0.5 & 0.5 \\ 0.5 & 0 \end{bmatrix}$$

Since its eigenvalues are $\lambda_1 = 0.8090$ and $\lambda_2 = -0.3090$, one gets $I_C = 0$.

Cauchy index of a symmetrical system can be also calculated from the signature matrix. It is given by the number of positive elements n_+ minus

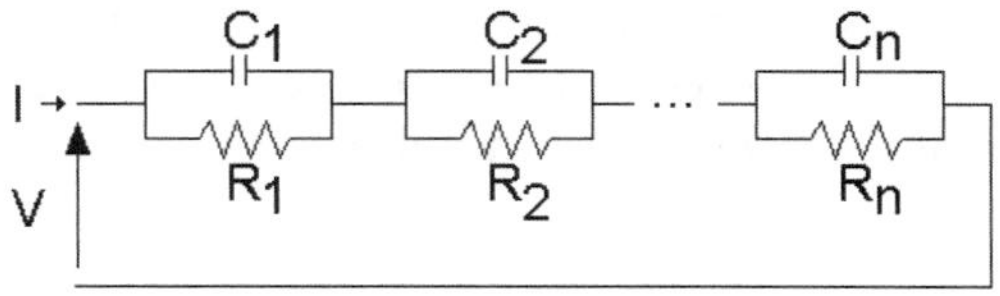

FIGURE 7.2
A relaxation circuit.

the negative elements n_- in the diagonal of the signature matrix, that is $I_C = n_+ - n_-$.

The Cauchy index is also linked to the difference between positive and negative residue numbers. In fact, if you consider a stable symmetrical SISO system in Jordan form with distinct and real eigenvalues, that is with $\mathrm{A} = \mathtt{diag}\{\lambda_1, \lambda_2, \ldots, \lambda_n\}$, $\mathrm{B}^T = \begin{bmatrix} 1 & 1 & \ldots & 1 \end{bmatrix}$ e $\mathrm{C} = \begin{bmatrix} R_1 & R_2 & \ldots & R_n \end{bmatrix}$, then

$$G(s) = \frac{R_1}{s - \lambda_1} + \frac{R_2}{s - \lambda_2} + \ldots + \frac{R_n}{s - \lambda_n}$$

and the Cauchy index is actually equal to the difference between the number of positive and negative residues.

When the Cauchy index equals the system order, that is when S = I, all the residues are positive and this is a *relaxation system*. In fact, the impulse response $y(t) = R_1 e^{-\lambda_1 t} + R_2 e^{-\lambda_2 t} + \ldots + R_n e^{-\lambda_n t}$ is given by the sum of all the positive terms which tend monotonically to zero, that is they relax toward zero. These systems have an exact electrical analogue in circuits which are series of elementary impedances, each formed by a resistor and a capacitor in parallel as in Figure 7.2. The generic block formed by a resistor R_i and a capacitor C_i in parallel has an impedance of $Z_i(s) = \frac{R_i}{sC_iR_i+1}$, whereas the impedance of the entire circuit is given by the sum of impedances of the single blocks: $Z(s) = Z_1(s) + Z_2(s) + \ldots + Z_n(s)$.

From theorem 11, it follows that relaxation systems have a balanced form of type $\mathrm{A} = \mathrm{A}^T$ and $\mathrm{B} = \mathrm{C}^T$.

Another property of the cross-gramian W_{co} is that it is a diagonal matrix in the reference system where the system is balanced. If S = I (so $\mathrm{A} = \mathrm{A}^T$ and $\mathrm{B} = \mathrm{C}^T$), it follows from the fact that equation $\mathrm{AW}_{co} + \mathrm{W}_{co}\mathrm{A} = -\mathrm{BC}$ equals $\mathrm{AW}_c^2 + \mathrm{W}_c^2\mathrm{A}^T = -\mathrm{BB}^T$ and $\mathrm{A}^T\mathrm{W}_o^2 + \mathrm{W}_o^2\mathrm{A} = -\mathrm{C}^T\mathrm{C}$, whereas in the more general case of $\mathrm{S} \neq \mathrm{I}$ this can be demonstrated by considering that in balanced form $\mathrm{W}_{co}^2 = \mathrm{W}_c^2\mathrm{W}_o^2 = \Sigma^2$.

From such considerations, a simplified procedure to obtain the balanced form for a symmetrical system can be derived. Obviously, the procedure in Chapter 5 can be applied but for symmetrical systems you can start by diagonalizing the cross-gramian W_{co}ordering the eigenvalues in decreasing order of absolute value. Thus, a reference system is obtained in which W_{co} is diagonal and in which the controllability and observability gramians are diagonal but not equal. At this point, proceed exactly as for example 5.2, to obtain the balanced form by rescaling the state variables.

MATLAB® exercise 7.3

Consider system $G(s) = \frac{s+2}{s^2+3s+5}$.
First, let us define the system:
```
>> s=tf('s')
>> G=(s+2)/(s^2+3*s+5)
```
and consider the canonical controllability form of the system
```
>> A=[0 1; -5 -3];
>> B=[0; 1];
>> C=[2 1];
>> system=ss(A,B,C,0)
```
Let us calculate the cross-gramian from Lyapunov's equation (7.8)
```
>> Wco=lyap(A,A,B*C)
```
and calculate the singular values of the system
```
>> eig(Wco)
```
the singular values of the system are $\sigma_1 = 0.2633$ and $\sigma_2 = 0.0633$.
It can be proven that the same result is obtained by calculating the singular values from the product of controllability and observability gramians:
```
>> Wc2=gram(system,'c')
>> Wo2=gram(system,'o')
>> sqrt(eig(Wc2*Wo2))
```
To calculate the balanced form, let us diagonalize the cross-gramian:
```
>> [T,L]=eig(Wco)
```
Matrix T is such that `inv(T)*Wco*T` is diagonal. Let us consider the system obtained applying the state transformation $\tilde{\mathbf{x}} = \mathrm{T}^{-1}\mathbf{x}$:
```
>> Atilde=inv(T)*A*T
>> Btilde=inv(T)*B
>> Ctilde=C*T
```
In this, the cross-gramian is diagonal:
```
>> Wcotilde=lyap(Atilde,Atilde,Btilde*Ctilde)
```
The gramians are also diagonal:
```
>> systemtilde=ss(Atilde,Btilde,Ctilde,0)
>> Wc2tilde=gram(systemtilde,'c')
>> Wo2tilde=gram(systemtilde,'o')
```
Note that this property derives from the fact that matrix T is diagonal:
```
>> Tsim=obsv(systemtilde)'*inv(ctrb(systemtilde))
```
So, now we can proceed as in example 5.2. However, it should be noted that since MATLAB®'s diagonalization procedure does not sequence the eigenvalues, the order of the two state variables must first be inverted using the transformation $\bar{x}_1 = \tilde{x}_2$ and $\bar{x}_2 = \tilde{x}_1$, or via the matrix:
```
>> Tbis=[0 1; 1 0];
>> Atilde=inv(Tbis)*Atilde*Tbis
>> Btilde=inv(Tbis)*Btilde
>> Ctilde=Ctilde*Tbis
```

7.5 Open-loop parameterization

The main issue with open-loop parameterization is determining which systems admit a certain set of assigned singular values. In the particular case of symmetrical SISO systems the issue is finding a system with an assigned signature matrix $S = \texttt{diag}\{s_1, s_2, \ldots, s_n\}$, a certain set of assigned singular values $\Sigma = \texttt{diag}\{\sigma_1, \sigma_2, \ldots, \sigma_n\}$ and assigned b_i coefficients of matrix $B^T = \begin{bmatrix} b_1 & b_2 & \ldots & b_n \end{bmatrix}$. Below, we will always assume that assigning singular values also means assigning a certain signature matrix.

Once the b_i coefficient and the signature matrix are known, matrix C is obtained from $C = B^T S$. As regards matrix A, the a_{ij} coefficients can be obtained from the following formula:

$$a_{ij} = -\{\frac{\sigma_i s_i s_j - \sigma_j}{\sigma_i^2 - \sigma_j^2}\} * BB^T \tag{7.11}$$

where symbol $*$ is the Hadamard product the component by component product of two square matrices, $F * G = \{f_{ij} g_{ij}\}$.

In particular, the diagonal terms of matrix A are given by the formula:

$$a_{ii} = -\frac{1}{2\sigma_i} b_i^2$$

Example 7.2

Construct a system with signature matrix $S = \begin{bmatrix} 1 & 0 & 0 \\ 0 & 1 & 0 \\ 0 & 0 & -1 \end{bmatrix}$, singular values σ_1, σ_2 e σ_3 and matrix $B = \begin{bmatrix} b_1 \\ b_2 \\ b_3 \end{bmatrix}$.

By applying the parameterization formula (7.11), we obtain: $a_{11} = -\frac{1}{2\sigma_1} b_1^2$, $a_{12} = -\frac{b_1 b_2}{\sigma_1 + \sigma_2}$, $a_{13} = -\frac{b_1 b_3}{\sigma_1 - \sigma_3}$, $a_{21} = -\frac{b_1 b_2}{\sigma_1 + \sigma_2}$, $a_{22} = -\frac{1}{2\sigma_2} b_2^2$, $a_{23} = -\frac{b_2 b_3}{\sigma_2 - \sigma_3}$, $a_{31} = -\frac{b_1 b_3}{\sigma_3 - \sigma_1}$, $a_{32} = -\frac{b_2 b_3}{\sigma_3 - \sigma_2}$, $a_{33} = -\frac{1}{2\sigma_3} b_3^2$.

Formula (7.11) can be obtained (to simplify consider the SISO case) from the gramians' equations expressed in the open-loop balanced realization: $W_o^2 = W_c^2 = \Sigma = \texttt{diag}\{\sigma_1, \sigma_2, \ldots, \sigma_n\}$:

$$\begin{aligned} A^T \Sigma + \Sigma A &= -C^T C \\ A \Sigma + \Sigma A^T &= -BB^T \end{aligned} \tag{7.12}$$

Summing the two matrix equations (7.12), we obtain

$$(A^T + A)\Sigma + \Sigma(A + A^T) = -C^T C - BB^T$$

and so

$$(A^T + A)\Sigma + \Sigma(A + A^T) = -SBB^T S - BB^T \tag{7.13}$$

The generic diagonal term of this expression is given by:

$$4a_{ii}\sigma_i = -2b_i^2$$

from which we obtain the expression which is valid for the diagonal terms of the parameterization (7.11).

Subtracting the second of (7.12) from the first:

$$(\mathrm{A}^T - \mathrm{A})\Sigma + \Sigma(\mathrm{A} - \mathrm{A}^T) = -\mathrm{C}^T\mathrm{C} + \mathrm{B}\mathrm{B}^T$$

and so

$$(\mathrm{A}^T - \mathrm{A})\Sigma + \Sigma(\mathrm{A} - \mathrm{A}^T) = -\mathrm{S}\mathrm{B}\mathrm{B}^T\mathrm{S} + \mathrm{B}\mathrm{B}^T \tag{7.14}$$

The diagonal terms of this expression are zero. The general terms a_{ij} of the parameterization (7.11) can be obtained from (7.13) and (7.14).

From equation (7.14) let us obtain the generic ij term:

$$(a_{ji} - a_{ij})\sigma_j + \sigma_i(a_{ij} - a_{ji}) = -s_i b_i b_j s_j + b_i b_j \tag{7.15}$$

whereas from relation (7.13) we obtain:

$$(a_{ji} + a_{ij})\sigma_j + \sigma_i(a_{ij} + a_{ji}) = -s_i b_i b_j s_j - b_i b_j \tag{7.16}$$

Equations (7.15) and (7.16) represent a system with two equations and two unknowns, a_{ij} and a_{ji}, which can be found by substitution obtaining a_{ij} from the relation which is the sum of the two prior relations (that is considering the generic term ij of the first equation (7.12)):

$$a_{ji} = \frac{-s_i s_j b_i b_j - \sigma_i a_{ij}}{\sigma_j} \tag{7.17}$$

Substituting in (7.15) we get:

$$-s_i s_j b_i b_j - \sigma_i a_{ij} - \sigma_j a_{ij} + \frac{\sigma_i}{\sigma_j}(s_i s_j b_i b_j + \sigma_i a_{ij}) + \sigma_i a_{ij} = -s_i s_j b_i b_j + b_i b_j \tag{7.18}$$

from which we obtain $a_{ij} = -\{\frac{\sigma_i s_i s_j - \sigma_j}{\sigma_i^2 - \sigma_j^2}\} * \mathrm{B}\mathrm{B}^T$.

Note that specifying a SISO system of order n through its transfer function $G(s)$, $2n$ parameters are required. Even for open-loop parameterization the assigned parameters are $2n$: n singular values and n matrix B coefficients.

If matrix B is not specified, there are an infinity of systems with matrix S and assigned singular values. To fix the system unequivocally, other n parameters must be specified, but they cannot just be any parameters. For example, it would not be possible to declare a priori if it is possible finding a system with n singular values and n assigned eigenvalues: such a problem could have a solution, several solutions or none at all. In fact, in the equations which could provide a solution (the unknowns are b_1, b_2, ..., b_n) we find nonlinear terms. By contrast, the solution to a similar problem has the dubious advantage of being a system with known stability properties (including the stability margins) and controllability and observability degrees.

Instead, notice that finding a system with $\mathrm{S} = \mathrm{I}$ and complex eigenvalues has no solution, $\mathrm{A} = \mathrm{A}^T$ being symmetrical and having real eigenvalues.

7.6 Relation between the Cauchy index and the Hankel matrix

Cauchy index I_c correlates with the properties of Hankel matrix H. Recall that Hankel matrix is defined as $\mathrm{H} = \mathrm{M}_o\mathrm{M}_c$ and is symmetrical by construction.

Suppose we have a SISO system with distinct eigenvalues and with this transfer function:

$$G(s) = \frac{\alpha_1}{s+\lambda_1} + \frac{\alpha_2}{s+\lambda_2} + \ldots + \frac{\alpha_n}{s+\lambda_n}$$

In this case, the Cauchy index can be calculated by counting the number of positive and negative residues: $I_c = N_{pos.res.} - N_{neg.res.}$. By highlighting the residue signs, each residue can be written as $\alpha_i = s_i|\alpha_i|$. Notice how this system can be associated with a Jordan form of type:

$$\mathrm{A} = \begin{bmatrix} -\lambda_1 & & & \\ & -\lambda_2 & & \\ & & \ddots & \\ & & & -\lambda_n \end{bmatrix}; \mathrm{B} = \begin{bmatrix} \sqrt{|\alpha_1|} \\ \sqrt{|\alpha_2|} \\ \vdots \\ \sqrt{|\alpha_n|} \end{bmatrix}; \mathrm{C}^T = \begin{bmatrix} s_1\sqrt{|\alpha_1|} \\ s_2\sqrt{|\alpha_2|} \\ \vdots \\ s_n\sqrt{|\alpha_n|} \end{bmatrix}$$

Since Hankel matrix H is:

$$\mathrm{H} = \begin{bmatrix} \mathrm{C} \\ \mathrm{CA} \\ \vdots \\ \mathrm{CA}^{n-1} \end{bmatrix} \begin{bmatrix} \mathrm{B} & \mathrm{AB} & \ldots \mathrm{A}^{n-1}\mathrm{B} \end{bmatrix} =$$

$$= \begin{bmatrix} \mathrm{B}^T\mathrm{S} \\ \mathrm{B}^T\mathrm{SA} \\ \vdots \\ \mathrm{B}^T\mathrm{SA}^{n-1} \end{bmatrix} \begin{bmatrix} \mathrm{B} & \mathrm{AB} & \ldots \mathrm{A}^{n-1}\mathrm{B} \end{bmatrix}$$

Given that A is diagonal then:

$$H = \begin{bmatrix} B^T S \\ B^T A^T S \\ \vdots \\ B^T (A^T)^{n-1} S \end{bmatrix} \begin{bmatrix} B & AB & \dots A^{n-1}B \end{bmatrix} = M_c^T S M_c$$

The Hankel matrix H is symmetrical. Whether it is defined positive or not depends exclusively on the signature matrix S. The signs of the eigenvalues (which are real) of the Hankel matrix reflect the number of positive and negative elements in the signature matrix. Furthermore, the Cauchy index can be calculated by counting the positive and negative eigenvalues in the Hankel matrix:

$$I_c = N_{\lambda_+(H)} - N_{\lambda_-(H)}$$

The Hankel matrix then helps understand the residue signs and the structure of the open-loop balanced form.

As we have already seen, for symmetrical systems $W_{co}^2 = W_c^2 W_o^2$ can be defined.

In this case, the hypothesis of asymptotic stability, normally necessary for non-symmetrical systems to deal with open-loop balancing, can be removed. Here however, it is no longer guaranteed that the eigenvalues of W_{co}^2 are real and positive. This condition in asymptotically stable minimal systems is assured by the fact that W_c^2 and W_o^2 matrices are positive definite. For symmetrical systems without the hypothesis of asymptotic stability, W_{co}^2 can even have negative or complex and conjugated eigenvalues. When, having calculated W_{co}^2, real and positive eigenvalues are found, then the singular values of the system can be defined as $\sigma_i = \sqrt{\lambda_i(W_{co}^2)}$.

It can be demonstrated that if the Hankel matrix is positive definite (i.e., if $I_c = n$ or $S = I$), then its eigenvalues $\lambda_i(W_{co}^2)$ are real and positive. Therefore, for non-asymptotically stable symmetrical systems, with a Hankel matrix positive definite, an open-loop balanced form can be defined. Clearly, the case where matrix A has purely imaginary eigenvalues should a priori be excluded since there is no solution to the Lyapunov equations and so a balanced form cannot be written.

7.7 Singular values for a FIR filter

Let us consider now discrete-time SISO systems. The transfer function $G(z) = \frac{Y(z)}{U(z)}$, as it is known, is the ratio of two polynomials of z. Two cases exist. If $G(z)$ can be written as $G(z) = h_1 z^{-1} + h_2 z^{-2} + \ldots + h_n z^{-n}$, that is as the sum of a finite number of terms of type $\frac{h_i}{z^i}$, then the system is said to be

FIR. Otherwise, if $G(z) = \frac{a_1 z^{-1} + a_2 z^{-2} + \ldots + a_{n-1} z^{-n+1}}{1 + b_1 z^{-1} + b_2 z^{-2} + \ldots + b_n z^{-n}}$, then the system is IIR (infinite impulse response).

FIR filters are distinguished by having all their poles at the origin so their transfer function is $G(z) = \frac{p(z)}{z^n}$ where $p(z)$ is an $n1$ order polynomial. Therefore FIR systems are asymptotically stable.

Furthermore, their name derives from the fact that they have an impulse response which cancels itself out after a finite number (n) of samples In fact, the output of a FIR system can be calculated from its transfer function:

$$Y(z) = G(z)U(z) = (h_1 z^{-1} + h_2 z^{-2} + \ldots + h_n z^{-n})U(z)$$

which is therefore given by

$$y(k) = h_1 u(k-1) + h_2 u(k-2) + \ldots + h_n u(k-n) \tag{7.19}$$

As you can see from the formula (7.19), if an input impulse is applied ($U(z) = 1$, i.e., $u(0) = 1$ and $u(k) = 0 \ \forall k \neq 0$), then the output will become null at sample $k = n + 1$, that is after n steps. The formula (7.19) also helps understand another important property of this class of systems: FIR systems have finite memory. In fact, output depends entirely on input regression so it is unnecessary to know system state to calculate the output at time k. This does not happen in IIR systems where output can be expressed by a model which includes regressions of the output itself:

$$y(k) = -b_1 y(k-1) - b_2 y(k-2) - \ldots - b_n y(k-n) + a_1 u(k-1) + \ldots + a_n u(k-n)$$

Therefore, to calculate the value of the output at time k in IIR systems, n prior samples of the output itself need to be memorized.

FIR filters do not bring about phase distortion. In continuous-time systems, as long as there is no phase distortion, the propagation time must be the same for all frequencies, so Bode diagram relative to phases must show a linear behavior with a steady decrease. With all their poles coinciding at the origin, FIR systems show in the discrete-time domain an analogous behavior.

The fundamental parameters of a FIR filter are $h_1, h_2, \ldots, h_n$, which can be found in the Hankel matrix which has a finite number of non-zero coefficients:

$$\mathrm{H} = \begin{bmatrix} h_1 & h_2 & \ldots & h_{n-1} & h_n \\ h_2 & h_3 & \ldots & h_n & 0 \\ h_3 & h_4 & \ldots & 0 & 0 \\ \vdots & & & & \vdots \\ h_n & 0 & \ldots & 0 & 0 \end{bmatrix}$$

Example 7.3

FIR filter $G(z) = z^{-1} + 5z^{-2}$ has Hankel matrix

$$\mathrm{H} = \begin{bmatrix} 1 & 5 \\ 5 & 0 \end{bmatrix}$$

Example 7.4

FIR filter $G(z) = z^{-1} + 5z^{-2} + 3z^{-3}$ has Hankel matrix

$$\mathrm{H} = \begin{bmatrix} 1 & 5 & 3 \\ 5 & 3 & 0 \\ 3 & 0 & 0 \end{bmatrix}$$

To calculate the balanced form for a discrete-time system the bilinear transformation $z = \frac{1+s}{1-s}$ is applied to obtain an equivalent continuous-time system. The theory valid for continuous-time systems is then applied to the equivalent system.

For FIR filters the procedure can be simplified. The following theorem can give the singular values for a FIR filter without recourse to the equivalent system.

Theorem 12 *The singular values of a FIR filter are given by the eigenvalues in absolute value of Hankel matrix* H.

Since the Hankel matrix H for a FIR filter is symmetrical, its eigenvalues are always real and their absolute value can be always computed.

Example 7.5

System $G(z) = h_2 z^{-2}$ is a second order FIR filter whose Hankel matrix is

$$\mathrm{H} = \begin{bmatrix} 0 & h_2 \\ h_2 & 0 \end{bmatrix}$$

The characteristic polynomial of matrix H is $p(\lambda) = \lambda^2 - h_2^2$, from which we obtain $\sigma_1 = \sigma_2 = |h_2|$.

Example 7.6

Now let us calculate the singular values of a FIR filter with transfer function $G(z) = h_3 z^{-3}$. The Hankel matrix of this system is:

$$\mathrm{H} = \begin{bmatrix} 0 & 0 & h_3 \\ 0 & h_3 & 0 \\ h_3 & 0 & 0 \end{bmatrix}$$

and singular values of the system are $\sigma_1 = \sigma_2 = \sigma_3 = |h_3|$.

Having presented several examples on singular values for a FIR filter, we will demonstrate theorem 12, which requires using the property that, given two matrices A and B, the eigenvalues of their product AB are equal to those of matrix BA: $\lambda_i(\mathrm{AB}) = \lambda_i(\mathrm{BA})$. In fact:

$$\det(\lambda\mathrm{I} - \mathrm{AB}) = \det(\mathrm{A}(\mathrm{A}^{-1}\lambda\mathrm{I} - \mathrm{B})) = \det(\mathrm{A})\det(\mathrm{A}^{-1}\lambda\mathrm{I} - \mathrm{B}) =$$

$$= \det(\mathrm{A}^{-1}\lambda\mathrm{I} - \mathrm{B})\det(\mathrm{A}) = \det((\mathrm{A}^{-1}\lambda\mathrm{I} - \mathrm{B})\mathrm{A}) = \det(\lambda\mathrm{I} - \mathrm{BA})$$

On this premise, let us consider the gramians of a FIR filter. The generic expression of gramians in discrete-time systems, reported in equations (5.18) and (5.19), for FIR filters is a finite sum:

$$W_c^2 = \sum_{i=0}^{n-1} A^i BB^T (A^T)^i \tag{7.20}$$

$$W_o^2 = \sum_{i=0}^{n-1} (A^T)^i C^T CA^i \tag{7.21}$$

and therefore

$$W_c^2 = BB^T + ABB^T A^T + \ldots + A^{n-1}BB^T(A^T)^{n-1} = M_c M_c^T \tag{7.22}$$

$$W_o^2 = C^T C + A^T C^T CA + \ldots + (A^T)^{n-1}C^T CA^{n-1} = M_o^T M_o \tag{7.23}$$

which follows that

$$\lambda_i(W_c^2 W_o^2) = \lambda_i(M_c M_c^T M_o^T M_o)$$

But, for the eigenvalue property of the product matrix:

$$\lambda_i(W_c^2 W_o^2) = \lambda_i(M_c^T M_o^T M_o M_c)$$

and therefore, given that $H = M_o M_c$, then

$$\lambda_i(W_c^2 W_o^2) = \lambda_i(H^T H)$$

from which it is clear that the singular values of a FIR filter are given by the eigenvalues in absolute value of Hankel matrix.

7.8 Singular values of all-pass systems

Let us consider a continuous-time system with transfer function $G(s) = k\frac{(5-s)^3}{(5+s)^3}$. The frequency response $G(j\omega)$ of this system has a very particular behavior. In fact, $\forall\omega\ |G(j\omega)| = 1$, and the magnitude Bode diagram of this system is flat.

All systems of this type are all-pass systems because the only difference between the input and output signals is phase, whereas the input amplitude is neither attenuated nor amplified. These systems are stable (all their poles are in the closed left-hand half of the complex plane), improper ($G(s) = D + \overline{G}(s)$), and not minimum phase (each pole has a zero symmetrical with respect to the imaginary axis and so with positive real part).

The singular values for these systems can also be calculated with ease.

Theorem 13 *The singular values for an all-pass system* $G(s) = k\frac{(a-s)^n}{(a+s)^n}$ *with* $k > 0$ *and* $a > 0$ *are* $\sigma_1 = \sigma_2 = \ldots = \sigma_n = k$.

This result derives from the fact that a bilinear transformation as well as transformations of type $\overline{s} = \frac{s}{a}$ do not change the singular values of a system. Bearing in mind these two considerations, and applying to $G(z) = kz^{-n}$ (whose singular values are $\sigma_1 = \ldots = \sigma_n = k$) at first the bilinear transformation and then a transformation of type $\overline{s} = \frac{s}{a}$, the result expressed by the theorem is obtained.

The importance of this result is that reduced order modeling cannot be applied to an all-pass system since all its singular values are equal.

MATLAB® exercise 7.4

Consider the system with transfer function:

$$G(s) = \frac{25s^4 - 352.5s^3 + 1830s^2 - 4120s + 3360}{s^4 + 14.1s^3 + 73.2s^2 + 164.8s + 134.4} \tag{7.24}$$

Let's define the system in MATLAB:

```
>> s=tf('s')
>> G=(25*s^4-352.5*s^3+1830*s^2-4120*s+3360)/(s^4+14.1*s^3+73.2*s^2+164.8*s+134.4)
```

The transfer function can be factorized as follows

```
>> zpk(G)
```

to obtain:

$$G(s) = 25\frac{(4-s)^3(2.1-s)}{(4+s)^3(2.1+s)} \tag{7.25}$$

which clearly shows that the system is all-pass with DC gain equal to $k = 25$.
Calculating the singular values of the system with the command:

```
>> [systembal,S]=balreal(G)
```

one obtains: $\sigma_1 = \sigma_2 = \sigma_3 = \sigma_4 = 25$.

7.9 Exercises

1. Determine, if possible, the system with eigenvalues $\lambda_1 = -1$, $\lambda_2 = -2$ and with singular values $\sigma_1 = 5$ and $\sigma_2 = 2$.
2. Calculate analytically the Cauchy index of the system $G(s) = \frac{s+1}{s(s^2+s+1)}$.
3. Given the system in Figure 7.3 with $R_1 = R_2 = 1$, $C_1 = 1$ and $C_2 = \frac{1}{2}$ determine the transfer function and the Cauchy index.
4. Write down an example of a relaxation system and verify the value of the Cauchy index.
5. Calculate the singular values of system $G(z) = \frac{3+2z+5z^2+6z^3}{z^4}$.

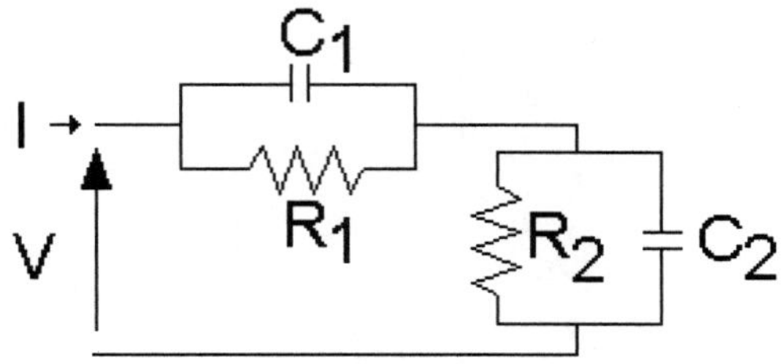

FIGURE 7.3
Circuit for exercise 3.

8

Linear quadratic optimal control

CONTENTS

In previous chapters using the right techniques, a lower order model of a system can be achieved. Designing the compensator for a high order system highlights the problem of approximating it which is examined in the next chapter. Before that, this chapter introduces a technique for synthesizing a linear quadratic regulator (LQR) which is able to determine the closed-loop eigenvalues on the basis of an optimized criteria. This design technique is the optimal control.

8.1 LQR optimal control

Let us consider a time-continuous system

$$\begin{aligned} \dot{\mathbf{x}} &= \mathrm{A}\mathbf{x} + \mathrm{B}\mathbf{u} \\ \mathbf{y} &= \mathrm{C}\mathbf{x} \end{aligned} \tag{8.1}$$

and let us suppose it is completely controllable and observable (alternatively consider only the part of the system which is controllable and observable). Using the control law $\mathbf{u} = -\mathrm{K}\mathbf{x}$, the system dynamic is governed by:

$$\dot{\mathbf{x}} = (\mathrm{A} - \mathrm{BK})\mathbf{x} \tag{8.2}$$

Usually, $\mathrm{A}_c = (\mathrm{A} - \mathrm{BK})$ indicates the closed-loop matrix of the system. As we know, the issue of state linear regulators is in selecting the right K gains so as to arbitrarily fix the A_c eigenvalues. Under the hypothesis of an observable and controllable system, there is always a solution.

Clearly, there are infinite ways to choose these eigenvalues. One is on the basis of criteria which include various specifications which a closed-loop system must satisfy. Once the criteria is defined, the choice of eigenvalues and so the K gains must be done to optimize the adopted criteria.

In the case of $\mathbb{L}_2$ or $\mathbb{H}_2$ optimal control the index (to minimize) is defined by the functional:

$$J = \int_0^\infty (\mathbf{x}^T \mathrm{Q}\mathbf{x} + \mathbf{u}^T \mathrm{R}\mathbf{u})dt \tag{8.3}$$

where the matrices $\mathrm{Q} \in \mathbb{R}^{n \times n}$ and $\mathrm{R} \in \mathbb{R}^{m \times m}$ are fixed weight matrices which define, as we will shortly see, the specifications for optimal control.

The optimal control issue lies in finding the K_{ott} gains so as to minimize functional (8.3).

As regards weight matrices, Q and R are generally positive definite. Q can also be defined as semi-definite positive, but R must always be defined positive.

The physical meaning of optimal control can be clarified through an example where the matrix Q is chosen in a particular way. Consider $\mathbf{z}(t) = \mathrm{C}_1\mathbf{x}(t)$ and the functional

$$J = \int_0^\infty (\mathbf{z}^T \mathbf{z} + \mathbf{u}^T \mathrm{R}\mathbf{u})dt$$

obtained from functional (8.3) making $\mathrm{Q} = \mathrm{C}_1^T\mathrm{C}_1$. Variables $\mathbf{z}(t)$ are not system output but other variables which we want to keep small. In other words, the control objective is to enable the $\mathbf{z}(t)$ variables to be very close to system equilibrium point ($\mathbf{x} = 0$ and so $\mathbf{z} = 0$). Supposing the system is initially excited (this is reflected by an initial condition $\mathbf{x}_0$ which is an undesired deviation from the equilibrium position $\mathbf{x} = 0$), the objective of the optimal control is finding the input so that the system most rapidly reaches equilibrium. As soon as the system is supposed controllable, this objective can always be obtained. Reaching equilibrium in the shortest possible time generally requires a considerable control signal which is unacceptable from certain points of view. Firstly, a considerable control signal could saturate the real system. Secondly, it could provoke a non-modeled high frequency dynamic in the original system. Therefore, two different costs need balancing $\mathbf{z}^T(t)\mathbf{z}(t) \geq 0$ and $\mathbf{u}^T(t)\mathrm{R}\mathbf{u}(t) > 0$ $\forall t$.

The significance of matrices Q and R can be therefore clarified by considering that the functional J represents the energy to minimize (for this reason it is often called the quadratic index). The functional takes into account the two energy terms: the first $J_1 = \int_0^\infty \mathbf{x}^T \mathrm{Q}\mathbf{x}dt$ deals with the energy associated with the closed-loop state variables weighted by matrix Q, whereas the other deals with the energy associated with input $\mathbf{u}$ and weighted by matrix R.

Obviously the choice of the closed-loop eigenvalues has to guarantee that the system is always asymptotically stable so that the zero-input response of the system tends to zero ($\mathbf{x}(t) \to 0$). From this consideration, minimizing functional $J_1 = \int_0^\infty \mathbf{x}^T \mathrm{Q}\mathbf{x}dt$ means making sure the state variables tend to zero as soon as possible. The smaller the energy associated with the state variables the more rapidly they tend to zero.

Minimizing the energy means ensuring a rapid transitory. This happens at the cost of input energy as we will see in this simple example.

Example 8.1

Consider $G(s) = \frac{1}{s+1}$. The unit-step response tends asymptotically to a value of one. Suppose output $y(t) = 1$ in the briefest time possible. By applying a step input with an amplitude of 10, the steady-state output value is exactly ten, and given the initial null conditions, this means that $y(t) = 1$ verified for a given t, less than for the previous case (unitary step input). In the limit case, choosing as input a Dirac impulse $u(t) = \delta(t)$, you obtain $y(t) = e^{-t}$ (impulse response) which is one for $t = 0$. So, at the cost of growing input energy, it is possible to reduce the time required for the output to reach a determined value.

Generally, the input energy required to obtain certain specifications for a closed-loop system should always be evaluated. For this reason, the quadratic index (8.3) also accounts for the input energy. Matrix R therefore assigns a relative weight to the two energy terms dealt with by the quadratic index. Matrix R weights the input therefore establishing if according to the control objective it is more important to minimize the first or second contribution. Matrix Q establishes the weight of the state variables, taking also into account for instance that it is not given that all the state variables have the same scale factor in the measures.

Now, let us see how the K_{ott} gains are determined. The K_{ott} gains are calculated from the following matrix equation:

$$\mathrm{PA} + \mathrm{A}^T\mathrm{P} - \mathrm{PBR}^{-1}\mathrm{B}^T\mathrm{P} + \mathrm{Q} = 0 \tag{8.4}$$

with $\mathrm{P} \in \mathbb{R}^{n \times n}$.

If $\overline{\mathrm{P}}$ is the solution, the optimum gains are given by:

$$\mathrm{K}_{ott} = \mathrm{R}^{-1}\mathrm{B}^T\overline{\mathrm{P}} \tag{8.5}$$

Note that there is always the inverse of R, in quanto R since it is a positive definite matrix.

The matrix equation (8.4), by contrast to the Lyapunov equations, is a nonlinear equation in P. In fact, a quadratic term ($\mathrm{PBR}^{-1}\mathrm{B}^T\mathrm{P}$) appears in the equation.

This equation is called the algebraic Riccati equation from the name of Jacopo Francesco, the Count of Riccati (1676–1754), who was the first to study equations of this type. The equation is known as algebraic because matrix P does not depend on time. There is a differential Riccati equation which is a function of time and it comes into play when, rather than defining the integral of the J index between zero and infinity, a finite horizon $[t_1, t_2]$ is considered in which to reach the control objective.

For systems with only one input (R is a scalar quantity) the quadratic term has a weight which is inversely proportional to increasing R.

Generally, the Riccati equation does not have a single solution. For example, if $n = 1$, it produces a second order equation with two solutions. Among the possible solutions for Riccati equations however, there is only one positive definite. It is this matrix $\overline{\mathrm{P}}$ which will provide the optimal gains K_{ott}.

Optimal gains have another fundamental property: they can guarantee closed-loop stability of the system, in other words $\mathrm{A}_c = \mathrm{A} - \mathrm{BK}_{ott}$ has all eigenvalues with negative real part.

To demonstrate that the K_{ott} gain guarantees closed-loop system stability, let us consider Riccati equation (8.4) with $\mathrm{P} = \overline{\mathrm{P}}$ and add and subtract $\overline{\mathrm{P}}\mathrm{BR}^{-1}\mathrm{B}^T\overline{\mathrm{P}}$:

$$\overline{\mathrm{P}}\mathrm{A} + \mathrm{A}^T\overline{\mathrm{P}} - \overline{\mathrm{P}}\mathrm{BR}^{-1}\mathrm{B}^T\overline{\mathrm{P}} + \overline{\mathrm{P}}\mathrm{BR}^{-1}\mathrm{B}^T\overline{\mathrm{P}} - \overline{\mathrm{P}}\mathrm{BR}^{-1}\mathrm{B}^T\overline{\mathrm{P}} + \mathrm{Q} = 0$$

Rearranging we obtain:

$$\overline{\mathrm{P}}(\mathrm{A} - \mathrm{BR}^{-1}\mathrm{B}^T\overline{\mathrm{P}}) + (\mathrm{A}^T - \overline{\mathrm{P}}\mathrm{BR}^{-1}\mathrm{B}^T)\overline{\mathrm{P}} + \overline{\mathrm{P}}\mathrm{BR}^{-1}\mathrm{B}^T\overline{\mathrm{P}} + \mathrm{Q} = 0$$

Considering $\mathrm{K}_{ott} = \mathrm{R}^{-1}\mathrm{B}^T\overline{\mathrm{P}}$, then

$$\overline{\mathrm{P}}(\mathrm{A} - \mathrm{BK}_{ott}) + (\mathrm{A}^T - \mathrm{K}_{ott}^T\mathrm{B}^T)\overline{\mathrm{P}} = -\mathrm{K}_{ott}^T\mathrm{RK}_{ott} - \mathrm{Q}$$

and so

$$\overline{\mathrm{P}}\mathrm{A}_c + \mathrm{A}_c^T\overline{\mathrm{P}} = -\mathrm{K}_{ott}^T\mathrm{RK}_{ott} - \mathrm{Q}$$

Since the right-hand term is a positive definite matrix because Q is positive semi-definite (or definite) and $\mathrm{K}_{ott}^T\mathrm{RK}_{ott}$ is positive definite, then the closed-loop system satisfies the Lyapunov equation for stability and according to Lyapunov second theorem it is asymptotically stable.

Where the Lyapunov equation can be solved in a closed form, calculating P requires an iterative equation. The iterative algorithm is based on the Kleinman method. The method is based on the property that the K gains are stabilizing (so they satisfy the Lyapunov equation) and on the relation linking K and P ($\mathrm{K} = \mathrm{R}^{-1}\mathrm{B}^T\mathrm{P}$).

In the Kleinman method, at the first iteration K_1 is fixed so that $\mathrm{A} - \mathrm{BK}_1$ is stable (if the closed-loop system with state matrix A is already stable, then $\mathrm{K}_1 = 0$). For generic iterations K_i is fixed by $\mathrm{K}_i = \mathrm{R}^{-1}\mathrm{B}^T\mathrm{P}_{i-1}$.

Once K_i is fixed, P_i is obtained by resolving the linear Lyapunov equation:

$$\mathrm{P}_i(\mathrm{A} - \mathrm{BK}_i) + (\mathrm{A}^T - \mathrm{K}_i^T\mathrm{B}^T)\mathrm{P}_i = -\mathrm{K}_i^T\mathrm{RK}_i - \mathrm{Q} \tag{8.6}$$

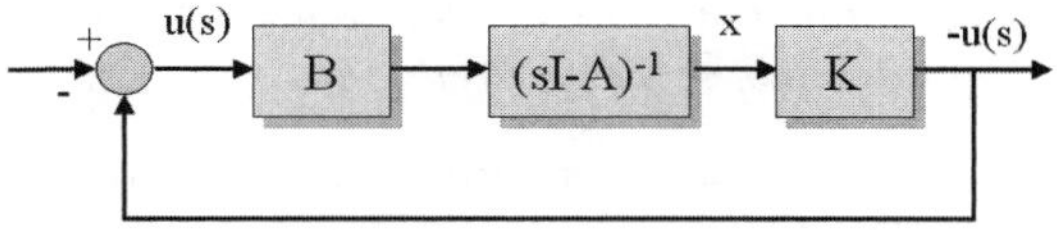

FIGURE 8.1
Block scheme of the linear quadratic regulator.

Next, the K_{i+1} gain is obtained by $K_{i+1} = R^{-1}B^TP_i$ iterating the procedure. The algorithm converges when $K_{i+1} \simeq K_i$.

Kleinman showed that starting from matrix K which guarantees closed-loop stability (matrix A_c) and iterating the procedure, the result is always a stable matrix so the various P_i matrices obtained are always definite positive matrices. The procedure converges on the definite positive solution of Riccati equation: $P_i \simeq P_{i+1} = \overline{P}$. Moreover, Kleinman showed that the convergence of this method is monotonic, that is the error norm at each step decreases monotonically. If instead, at the first iteration K_1 is not such that the closed-loop system with state matrix A_c is asymptotically stable, the method, given it converges, converges to a non-definite matrix.

The obvious advantage of the method is resolving Lyapunov equations iteratively to solve an algebraic Riccati equation, that is solving linear equations with closed solutions instead of nonlinear ones.

It can be shown that index J when $\mathbf{u} = -K_{ott}\mathbf{x}$ is $J = \frac{1}{2}\mathbf{x}^T(0)\overline{P}\mathbf{x}(0)$, which is therefore the smallest value the index can have $\forall\ \mathbf{x}(0)$.

Optimal control has however some disadvantages. First, it is based on feedback of all the state variables $\mathbf{x}$. This requires that the whole state is available for feedback. Optimal control therefore requires a sensor for every state variable, or it requires an observer to re-construct the variables which are not directly measurable. Optimal control is difficult to apply to flexible systems which require a higher order model (if not infinite).

Another disadvantage with optimal control is the gap with classical control specifics such as reaction to disturbance, over-elongation, stability margins, and so on. This often leads to a process of trial and error in defining the weight matrices R and Q.

Figure 8.1 shows the block scheme of optimal control (also called linear quadratic regulator). The transfer matrix is given by $G_{LQ} = K(sI - A)^{-1}B$. It can be shown that this matrix has certain robustness margins with respect to delays and gains in the direct chain.

The main features of optimal control are summarized in the following theorem:

Theorem 14 *Given system* $\dot{\mathbf{x}} = A\mathbf{x} + B\mathbf{u}$ *with initial condition* $\mathbf{x}_0$ *and given the index* $J = \int_0^\infty (\mathbf{x}^TQ\mathbf{x} + \mathbf{u}^TR\mathbf{u})dt$ *if the system is minimal, the entire* *state* $\mathbf{x}$ *can be fed back and matrices* Q *and* R *are symmetrical and positive semi-definite and definite, respectively, then:*

1. there is only one linear quadratic regulator $\mathbf{u} = -K_{ott}x$ *(with* $K_{ott} = R^{-1}B^T\bar{P}$*) which minimizes index* J*;*

2. $\bar{P}$ *is the only symmetrical positive definite solution of Riccati equation*

$$PA + A^TP - PBR^{-1}B^TP + Q = 0$$

3. the closed-loop system with state matrix $(A - BK_{ott})$ *is asymptotically stable;*

4. the minimum value of index J *is* $J = \frac{1}{2}\mathbf{x}_0^T\bar{P}\mathbf{x}_0$.

MATLAB® exercise 8.1

Here we discuss the use of the Kleinman algorithm for resolving the Riccati equation. Let us consider system $G(s) = \frac{s+2}{s^2-2s-3}$ which is stable and minimal. Consider the functional (8.3) with $Q = C^TC$ and $r = 1$. Let us define the system in MATLAB® from its canonical controllability form:

```
>> A=[0 1; 3 2];
>> B=[0; 1];
>> C=[2 1];
>> D=0;
```

Define the weight matrices

```
>> R=1;
>> Q=C'*C;
```

Choose K_0 such that $A_c = A - BK_0$ is asymptotically stable, and in particular that its eigenvalues are $\lambda_1 = -1$ e $\lambda_2 = -0.5$

```
>> K0=acker(A,B,[-1 -0.5]);
>> Ki=K0;
```

Apply the Kleinman method assuming that $K_{i+1} \simeq K_i$, when $\|K_{i+1} - K_i\| < 0.0001$:

```
>> for i=1:100
     P=lyap((A-B*Ki)',Ki'*R*Ki+Q);
     Kii=inv(R)*B'*P;
     if (norm(Kii-Ki)<0.0001), break; end
     Ki=Kii;
   end
```

After seven iterations the algorithm converges on $K_i = \begin{bmatrix} 6.6056 & 6.2674 \end{bmatrix}$ and $P = \begin{bmatrix} 7.3865 & 6.6056 \\ 6.6056 & 6.2674 \end{bmatrix}$. It is easy to verify that P is positive definite. The optimal eigenvalues (`eig(A-B*Ki)`) are $\lambda_{1,ott} = -1.1605$ and $\lambda_{2,ott} = -3.1070$.

Note that from unstable matrix A_c (e.g., by imposing eigenvalues $\lambda_1 = -1$ and $\lambda_2 = 2$), the algorithm converges to a non-definite matrix P.

The linear quadratic regulator can be also calculated with the MATLAB command:

```
>> [K,P,E]=lqr(A,B,C'*C,1)
```

Note that the same results are obtained.

Finally, the transfer function G_{LQ} is calculated with command:

```
>> system=ss(A,B,K,0)
```

To verify that the closed-loop system eigenvalues are effectively the optimal ones, the closed-loop transfer function can be calculated with command `>> tf(feedback(system,1))`. Furthermore, the Nyquist diagram can be plotted to verify the robustness of the optimal control in terms of gain and phase margins (command `>> nyquist(system)`).

MATLAB® exercise 8.2

Consider now the system with transfer function $G(s) = \frac{s+10}{s^2+s-2}$. As in the previous example, $G(s)$ is unstable and in minimal form. We want to calculate the linear quadratic regulator with $\mathrm{Q} = \mathrm{C}^T\mathrm{C}$ for two cases: $r = 0.1$ and $r = 10$.

Define the system in MATLAB by considering its canonical control form:

```
>> A=[0 1; 2 -1];
>> B=[0; 1];
>> C=[10 1];
>> D=0;
```

Then define the weight matrices:

```
>> Q=C'*C;
>> r1=0.1;
>> r2=10;
```

The linear quadratic regulator transfer function G_{LQ} for the two cases can be calculated through the commands:

```
>> [K1,P1,E1]=lqr(A,B,Q,r1);
>> GLQ1=ss(A,B,K1,0);
>> [K2,P2,E2]=lqr(A,B,Q,r2);
>> GLQ2=ss(A,B,K2,0);
```

and the corresponding closed-loop transfer functions are:

```
>> CLsys1=tf(feedback(GLQ1,1));
>> CLsys2=tf(feedback(GLQ2,1));
```

In order to compare the behavior of the two closed-loop systems, we can calculate and plot the zero-input response for both cases:

```
>> [Y1,T1,X1] = lsim(CLsys1,zeros(5001,1),[0:0.001:5],[1 1]);
>> [Y2,T2,X2] = lsim(CLsys2,zeros(5001,1),[0:0.001:5],[1 1]);
```

In Figure 8.2 the trends of the state variables $x_1(t)$ and $x_2(t)$ are reported for the two cases. When $r = 0.1$ (continuous lines) the energy associated to the input is weighted less than that associated to the states, hence the effect of the linear quadratic regulator, which mostly minimizes the state energy, is a zero-input response with a small time constant which rapidly decreases to zero; on the contrary, when $r = 10$ (dashed lines) the closed-loop system has a higher time constant, as it can be easily observed from its zero-input response.

8.2 Hamiltonian matrices

We have seen how optimal control is solved by an algebraic Riccati equation (8.4). To each equation of this type a matrix can be associated, a Hamiltonian matrix, which has very particular properties. We can, for example, associate to Riccati equation (8.4), the Hamiltonian matrix:

$$\mathrm{H} = \begin{bmatrix} \mathrm{A} & -\mathrm{B}\mathrm{R}^{-1}\mathrm{B}^T \\ -\mathrm{Q} & -\mathrm{A}^T \end{bmatrix} \tag{8.7}$$

H is a $2n \times 2n$ matrix whose first line are the coefficients (with their sign) of the term which is multiplied to the left by the unknown matrix P (element H_{11}) by the quadratic (component H_{12}). The other two coefficients of the

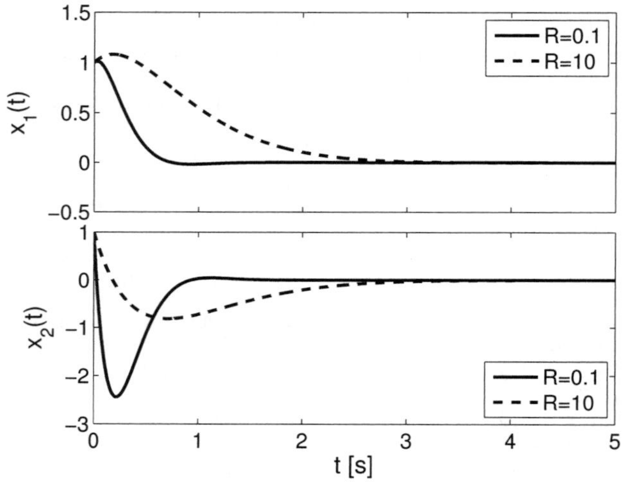

FIGURE 8.2
Zero-input response of the LQR closed-loop system for $r = 0.1$ and $r = 10$.

matrix (H_{21} and H_{22}) are given by the known term from the Riccati equation and by the term which is multiplied to the right by P, sign reversed.

Hamiltonian matrix H has prerogatives which, given

$$Y = \begin{bmatrix} 0 & I \\ -I & 0 \end{bmatrix}$$

with I the $n \times n$ identity matrix, then:

$$H^T Y = -YH \tag{8.8}$$

or rather

$$Y^{-1} H^T Y = -H \tag{8.9}$$

This is the property which defines the Hamiltonian matrix. All matrices with this property are called Hamiltonians.

The matrix associated with the Riccati equation for optimal control has a further important property expressed by the following theorem:

Theorem 15 *Given the optimal eigenvalues* $\lambda_1, \lambda_2, \ldots, \lambda_n$ *(i.e., the eigenvalues of* $A - BK_{ott}$*), the eigenvalues of* $H = \begin{bmatrix} A & -BR^{-1}B^T \\ -Q & -A^T \end{bmatrix}$ *are:*

$$\lambda_1, \lambda_2, \ldots, \lambda_n, -\lambda_1, -\lambda_2, \ldots, -\lambda_n$$

Proof 5 *Let us consider matrix* $T = \begin{bmatrix} I & 0 \\ \overline{P} & I \end{bmatrix}$ *and calculate its inverse* $T^{-1} = \begin{bmatrix} T_{11} & T_{12} \\ T_{21} & T_{22} \end{bmatrix}$ *(remembering that* $\overline{P}$ *is positive definite, so the inverse matrix of* T *exists), making* $T^{-1}T = I$.

Matrix $T^{-1}T$ *is given by:*

$$T^{-1}T = \begin{bmatrix} T_{11} & T_{12} \\ T_{21} & T_{22} \end{bmatrix} \begin{bmatrix} I & 0 \\ \overline{P} & I \end{bmatrix} = \begin{bmatrix} T_{11} + T_{12}\overline{P} & T_{12} \\ T_{21} + T_{22}\overline{P} & T_{22} \end{bmatrix}$$

Making $T^{-1}T = I$*, then:*

$$\begin{array}{l} T_{11} + T_{12}\overline{P} = I \Rightarrow T_{11} = I \\ T_{12} = 0 \\ T_{21} + T_{22}\overline{P} = 0 \Rightarrow T_{21} = -\overline{P} \\ T_{22} = I \end{array}$$

and so $T^{-1} = \begin{bmatrix} I & 0 \\ -\overline{P} & I \end{bmatrix}$.

Now let us consider matrix $T^{-1}HT$ *which, as we know, has the same eigenvalues of matrix* H*:*

$$T^{-1}HT = \begin{bmatrix} I & 0 \\ -\overline{P} & I \end{bmatrix} \begin{bmatrix} A & -BR^{-1}B^T \\ -Q & -A^T \end{bmatrix} \begin{bmatrix} I & 0 \\ \overline{P} & I \end{bmatrix} =$$

$$= \begin{bmatrix} A & -BR^{-1}B^T \\ -\overline{P}A - Q & \overline{P}BR^{-1}B^T - A^T \end{bmatrix} \begin{bmatrix} I & 0 \\ \overline{P} & I \end{bmatrix} =$$

$$= \begin{bmatrix} A - BR^{-1}B^T\overline{P} & -BR^{-1}B^T \\ -\overline{P}A - Q + \overline{P}BR^{-1}B^T\overline{P} - A^T\overline{P} & \overline{P}BR^{-1}B^T\overline{P} - A^T \end{bmatrix} =$$

$$= \begin{bmatrix} A - BK_{ott} & -BR^{-1}B^T \\ 0 & -(A - BK_{ott})^T \end{bmatrix}$$

In finding the last expression, the algebraic Riccati equation was used ($\overline{P}A + Q - \overline{P}BR^{-1}B^T\overline{P} + A^T\overline{P} = 0$*). Notice that the* H *eigenvalues are given by union of* $\lambda_1, \lambda_2, \ldots, \lambda_n$ *and of* $-\lambda_1, -\lambda_2, \ldots, -\lambda_n$.

For SISO systems, theorem 15 also provides an alternative method for finding the optimal controller to that based on Riccati equation (8.4). First, the matrix H eigenvalues are found and those on the right-hand half of the complex plane are discarded. Then, those eigenvalues with negative real part are used to design a control law $\mathbf{u} = -K\mathbf{x}$ by eigenvalue placement (a system of n equations with n unknowns). The method guarantees that no eigenvalues are found on the imaginary axis: if there are H eigenvalues on the imaginary axis the system is uncontrollable. If this were to be verified, the base hypothesis for solving optimal control would be violated. Furthermore in the SISO case, the Lyapunov equation could be used to find $\overline{\mathrm{P}}$: $\mathrm{A}_c^T\overline{\mathrm{P}} + \overline{\mathrm{P}}\mathrm{A}_c = -\mathrm{K}_{ott}^T\mathrm{R}\mathrm{K}_{ott} - \mathrm{Q}$.

The property illustrated is characteristic of all Hamiltonian matrices, all of which have symmetrical eigenvalues with respect to the imaginary axis, or rather if λ is an eigenvalue, then so too is $-\lambda$. However, the opposite is not true as the following example shows.

MATLAB® exercise 8.3

Consider the matrix $\mathrm{P} = \begin{bmatrix} 0 & 1 & 0 & 0 \\ 0 & 0 & 1 & 0 \\ 0 & 0 & 0 & 1 \\ 1 & 0 & 0 & 0 \end{bmatrix}$. Despite its eigenvalues $\lambda_{1,2} = \pm 1$ and $\lambda_{3,4} = \pm j$, it is not a Hamiltonian matrix. In fact, if we calculate $-\mathrm{Y}^{-1}\mathrm{P}^T\mathrm{Y}$, we obtain

$$-\mathrm{Y}^{-1}\mathrm{P}^T\mathrm{Y} = \begin{bmatrix} 0 & 0 & 0 & 1 \\ -1 & 0 & 0 & 0 \\ 0 & 1 & 0 & 0 \\ 0 & 0 & -1 & 0 \end{bmatrix}$$

i.e., $-\mathrm{Y}^{-1}\mathrm{P}^T\mathrm{Y} \neq \mathrm{P}$, and P does not satisfy equation (8.9). This can be checked in MATLAB with the following commands:

```
>> H=diag([1 1 1],1)
>> H(4,1)=1
>> eig(H)
>> Y=[zeros(2) eye(2); -eye(2) zeros(2)]
>> -inv(Y)*H'*Y
```

This example demonstrates that it is not true that, if a matrix has eigenvalues $\lambda_1, \lambda_2, \ldots, \lambda_n, -\lambda_1, -\lambda_2, \ldots, -\lambda_n$, it is Hamiltonian.

MATLAB® exercise 8.4

In this MATLAB® exercise we show an example of the calculation of the optimal eigenvalues from the Hamiltonian matrix. We then show how they can be assigned as closed-loop eigenvalues with the command `acker`.

Consider the continuous-time LTI system given by

$$\mathrm{A} = \begin{bmatrix} -1 & 0 & 0 \\ 0 & 2 & 0 \\ 0 & 0 & 3 \end{bmatrix}; \mathrm{B} = \begin{bmatrix} 1 \\ 1 \\ 1 \end{bmatrix}; \mathrm{C} = \begin{bmatrix} 1 & 1 & 1 \end{bmatrix} \tag{8.10}$$

and the optimal control problem with

$$\mathrm{Q} = \begin{bmatrix} 2 & 0 & 0 \\ 0 & 1 & 0 \\ 0 & 0 & 1 \end{bmatrix}; \mathrm{R} = 15 \tag{8.11}$$

Let us first define in MATLAB® the state-space matrices:

```
>> A=diag([-1 2 3])
>> B=[1; 1; 1]
>> C=[1 1 1]
```

and matrices Q and R:

```
>> R=15
>> Q=[2 0 0; 0 1 0; 0 0 1]
```

We then calculate the Hamiltonian matrix H and its eigenvalues:

```
>> H=[A -B*inv(R)*B'; -Q -A']
```

```
>> p=eig(H)
```

One gets: $p_1 = 3.0114$, $p_2 = 2.0171$, $p_3 = 1.0627$, $p_4 = -1.0627$, $p_5 = -2.0171$ and $p_6 = -3.0114$.

The optimal closed-loop eigenvalues are those with negative real part, i.e., p_4, p_5 and p_6. The value of the gain that corresponds to such eigenvalues can be calculated with the command `acker` as follows:

```
>> K=acker(A,B,p(4:6))
```

The result is a closed-loop system having as eigenvalues p_4, p_5 and p_6 as it can be verified with the following command:

```
>> eig(A-B*K)
```

8.3 Resolving the Riccati equation by Hamiltonian matrix

Riccati equation (8.4) can be solved by a method which holds true for all quadratic equations and is based on the associated Hamiltonian matrices. Consider a diagonalization of matrix H, obtained by ordering the Hamiltonian eigenvalues so that those at the top of the diagonal have negative real part:

$$\mathrm{H} = \mathrm{T}\begin{bmatrix} -\Lambda & 0 \\ 0 & \Lambda \end{bmatrix}\mathrm{T}^{-1} \tag{8.12}$$

with $\mathrm{T} = \begin{bmatrix} \mathrm{T}_{11} & \mathrm{T}_{12} \\ \mathrm{T}_{21} & \mathrm{T}_{22} \end{bmatrix}$. At this point P can be calculated:

$$\mathrm{P} = \mathrm{T}_{21}\mathrm{T}_{11}^{-1} \tag{8.13}$$

The eigenvalue order within sub-matrix Λ is irrelevant as regards this calculation. Ordering in terms of eigenvalues with negative real part and those with positive real part is possible only if matrix H has no imaginary eigenvalues, a hypothesis required for solving the Riccati equation.

This method based on Jordan decomposition of matrix H can prove to be computationally demanding. At the end of the 80s a method based on another decomposition (Schur decomposition) proved more advantageous.

In Schur decomposition, matrix H is the product of three matrices:

$$\mathrm{H} = \mathrm{USU}^T \tag{8.14}$$

with $\mathrm{S} = \begin{bmatrix} \mathrm{S}_{11} & \mathrm{S}_{12} \\ 0 & \mathrm{S}_{22} \end{bmatrix}$ and U an orthonormal matrix ($\mathrm{UU}^T = \mathrm{I}$). So, S is a higher triangular matrix or quasi-triangular. Furthermore, it is structured such that S_{11} has eigenvalues with negative real part, whereas S_{22} has eigenvalues with positive real part. The solution to the Riccati equation is found analogously. Since $\mathrm{U} = \begin{bmatrix} \mathrm{U}_{11} & \mathrm{U}_{12} \\ \mathrm{U}_{21} & \mathrm{U}_{22} \end{bmatrix}$, matrix P is given by:

$$P = U_{21}U_{11}^{-1} \tag{8.15}$$

8.4 The Control Algebraic Riccati Equation

Let us now consider a particular case of optimal control. Consider a linear time-invariant linear system:

$$\begin{aligned} \dot{\mathbf{x}} &= \mathrm{A}\mathbf{x} + \mathrm{B}\mathbf{u} \\ \mathbf{y} &= \mathrm{C}\mathbf{x} \end{aligned} \tag{8.16}$$

and an LQR problem with the following index:

$$J = \int_0^\infty (\mathbf{y}^T\mathbf{y} + \mathbf{u}^T\mathbf{u})dt \tag{8.17}$$

This index defined by equation (8.17) is a special case of the index defined by equation (8.3), from $Q = C^TC$ and $R = I$.

Note that in this case (8.17) has two terms: term $\int_0^\infty \mathbf{y}^T\mathbf{y}dt$ is the energy associated to the system output, whereas $\int_0^\infty \mathbf{u}^T\mathbf{u}dt$ is the energy associated to the system input.

Riccati equation (8.4) associated with this problem is called the Control Algebraic Riccati Equation (CARE):

$$A^TP + PA - PBB^TP + C^TC = 0 \tag{8.18}$$

The Hamiltonian matrix associated to the CARE is:

$$H = \begin{bmatrix} A & -BB^T \\ -C^TC & -A^T \end{bmatrix} \tag{8.19}$$

If the system is not strictly proper

$$\begin{aligned} \dot{\mathbf{x}} &= \mathrm{A}\mathbf{x} + \mathrm{B}\mathbf{u} \\ \mathbf{y} &= \mathrm{C}\mathbf{x} + \mathrm{D}\mathbf{u} \end{aligned} \tag{8.20}$$

the CARE equation becomes:

$$A^TP + PA - (PB + C^TD)(I + D^TD)^{-1}(B^TP + D^TC) + C^TC = 0 \tag{8.21}$$

and the Hamiltonian becomes:

$$H = \begin{bmatrix} A - B(I + D^TD)^{-1}D^TC & -B(I + D^TD)^{-1}B^T \\ -C^TC + C^TD(I + D^TD)^{-1}D^TC & -A^T + C^TD(I + D^TD)^{-1}B^T \end{bmatrix} \tag{8.22}$$

MATLAB® exercise 8.5

Consider the continuous-time LTI system with state-space matrices:

$$\mathrm{A}=\left[\begin{array}{cc} -1 & 0 \\ 0 & 3 \end{array}\right];\mathrm{B}=\left[\begin{array}{c} 1 \\ -1 \end{array}\right];\mathrm{C}=\left[\begin{array}{cc} 2 & 1 \end{array}\right];\mathrm{D}=1 \tag{8.23}$$

and let's compute the optimal control with index (8.17).
Once the system has been defined

```
>> A=[-1 0; 0 3]
>> B=[1; -1]
>> C=[2 1]
>> D=1
```

let's compute the Hamiltonian

```
>> H=[A-B*inv(eye(1)+D'*D)*D'*C -B*inv(eye(1)+D'*D)*B';
-C'*C+C'*D*inv(eye(1)+D'*D)*D'*C -A'+C'*D*inv(eye(1)+D'*D)*B']
```

and its eigenvalues

```
>> eig(H)
```

We find the optimal eigenvalues: $\lambda_1 = -3.1794$ and $\lambda_2 = -2.3220$.
It is also possible to use the `lqr` command as follows:

```
>> Q=C'*C-C'*D*inv(eye(1)+D'*D)*D'*C
>> [K,P,E]=lqr(A-B*inv(eye(1)+D'*D)*D'*C,B,Q,(eye(1)+D'*D))
```

8.5 Optimal control for SISO systems

Let's now consider the case of linear time-invariant SISO systems. For this class of systems there exist results of significant interest in simplifying optimal control. In particular, let's consider a generalization of index (8.17) so as to weight unequally the energy terms associated with the input and output:

$$J=\int_0^{\infty}(\mathbf{y}^T\mathbf{y}+\mathbf{u}^T r\mathbf{u})dt=\int_0^{\infty}(y^2+ru^2)dt \tag{8.24}$$

This optimal control problem is the same as that associated with (8.3) on condition that $\mathrm{Q}=\mathrm{C}^T\mathrm{C}$ and $\mathrm{R}=r$ (note that because it is a SISO system, $\mathrm{R}\in\mathbb{R}^{1\times 1}$ is a scalar quantity).

Let's suppose the system has transfer function $G(s)=\frac{b(s)}{a(s)}$ and it is minimal.

To resolve optimal control a realization (A, B, C) of system could be adopted and its optimal eigenvalues be found from the Hamiltonian:

$$\mathrm{H}=\left[\begin{array}{cc} \mathrm{A} & -\mathrm{B}r^{-1}\mathrm{B}^T \\ -\mathrm{C}^T\mathrm{C} & -\mathrm{A}^T \end{array}\right] \tag{8.25}$$

Let's remember that generally, given the weight matrices Q and R (in the SISO case, fixed r), the optimal eigenvalues do not depend on the chosen reference system. In this case, therefore, the optimal eigenvalues do not depend on the adopted system realization (A, B, C) but only on transfer function $G(s)$.

For SISO systems there is another method for finding optimum eigenvalues which is simpler. They are given by the Letov theorem.

Theorem 16 (Letov theorem) *For a minimal linear time-invariant SISO system with transfer function* $G(s) = \frac{b(s)}{a(s)}$*, the optimal eigenvalues according to index (8.24) are given by the roots with negative real part of the following equation:*

$$a(s)a(-s) + r^{-1}b(s)b(-s) = 0 \tag{8.26}$$

Example 8.2

Consider a linear time-invariant SISO system with transfer function $G(s) = \frac{1}{s}$. Suppose you wanted to find the optimal eigenvalues according to the functional (8.24) with $r = 1$. Applying the Letov theorem then

$$-s^2 + 1 = 0 \tag{8.27}$$

and so the optimal eigenvalue is $\lambda_{ott} = -1$.
Alternatively, a realization of the system is given by $\mathrm{A} = 0$, $\mathrm{B} = 1$, $\mathrm{C} = 1$, and the CARE equation is:

$$-\mathrm{P}^2 + 1 = 0 \Rightarrow \bar{\mathrm{P}} = 1 \tag{8.28}$$

So $\mathrm{K}_{ott} = -\mathrm{B}\bar{\mathrm{P}} = -1$. Finally, given that $\mathrm{A}_c = \mathrm{A} - \mathrm{BK}_{ott} = -1$, the optimal eigenvalue is $\lambda_{ott} = -1$.

Example 8.3

Let us consider the linear time-invariant SISO system with transfer function $G(s) = \frac{10-s}{(s+1)(s-2)}$ and suppose $r = 2$. In this case the optimal eigenvalues are given by:

$$(s+1)(s-2)(1-s)(-s-2) + \frac{1}{2}(10-s)(10+s) = 0$$

$$\Rightarrow s^4 - 5.5s^2 + 54 = 0 \tag{8.29}$$

The solutions to equation (8.29) are $s_{1,2} = -2.2471 \pm j1.5163$ and $s_{3,4} = 2.2471 \pm j1.5163$ (see Figure 8.3). The optimal eigenvalues are $\lambda_{1,2} = -2.2471 \pm j1.5163$.

Example 8.4

Let's consider a linear time-invariant SISO system with transfer function $G(s) = \frac{1}{s^2}$ and let $r = 1$. In this case the Letov formula becomes:

$$s^4 + 1 = 0 \tag{8.30}$$

The solutions to equation (8.30) are $s_{1,2} = -0.7071 \pm j0.7071$ and $s_{3,4} = 0.7071 \pm j0.7071$ and they belong to the circumference of unitary radius (Figure 8.4).

The result in example 8.4 can be generalized to a system of n integrators in cascade ($G(s) = \frac{1}{s^n}$). In fact, in this case ($r = 1$), the equation for obtaining the optimal eigenvalues becomes:

$$(-1)^n s^{2n} + 1 = 0 \tag{8.31}$$

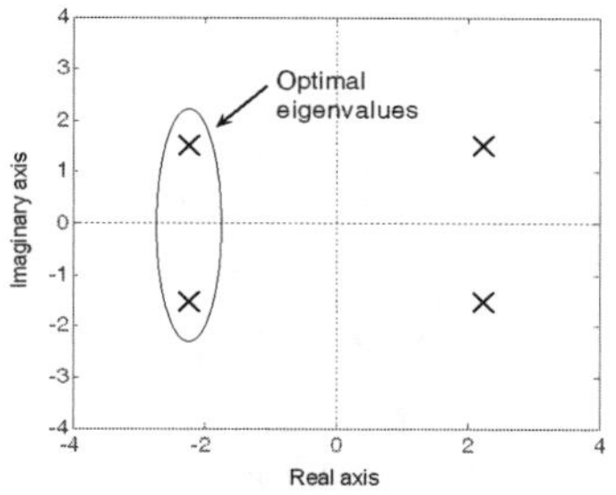

FIGURE 8.3
Solutions for equation (8.29) and optimal eigenvalues ($r = 2$) for the system $G(s) = \frac{10-s}{(s+1)(s-2)}$.

Example 8.5

Let's consider the all-pass system $G(s) = \frac{(1-s)(2-s)}{(1+s)(2+s)}$. In this case the Letov formula becomes:

$$(1-s)(2-s)(1+s)(2+s) + \frac{1}{r}(1-s)(2-s)(1+s)(2+s) = 0$$

$$\Rightarrow (1 + \frac{1}{r})(1-s)(2-s)(1+s)(2+s) = 0$$

The optimal eigenvalues are $\lambda_1 = -1$ and $\lambda_2 = -2$ $\forall r$.

Generalizing the result of the previous exercise, it may be intuited that for whatever type of all-pass system the optimal eigenvalues do not depend on r and are given by the eigenvalues (if it is the case, sign changed) of the open-loop system.

Example 8.6

$G(s) = \frac{(1+s)(2-s)}{(1-s)(2+s)}$ is an all-pass system. For $\forall r$ the optimal eigenvalues are $\lambda_1 = -1$ and $\lambda_2 = -2$.

Example 8.7

Consider now a system $G(s) = \frac{b(s)}{a(s)}$ satisfying the following property: $G(s) = -G^T(-s)$. For such a system the optimal eigenvalues for $r = 1$ are given by the roots of $a(s)+b(s) = 0$. In fact, since $\frac{b(s)}{a(s)} = -\frac{b(-s)}{a(-s)}$

$$b(s)a(-s) + b(-s)a(s) = 0 \tag{8.32}$$

and since from the Letov formula with $r = 1$ one has

$$a(s)a(-s) + b(s)b(-s) = 0 \tag{8.33}$$

adding equations (8.32) and (8.33) one gets:

$$a(s)a(-s) + b(s)b(-s) + b(s)a(-s) + b(-s)a(s) = 0$$

and thus

$$(a(-s) + b(-s))\,(a(s) + b(s)) = 0$$

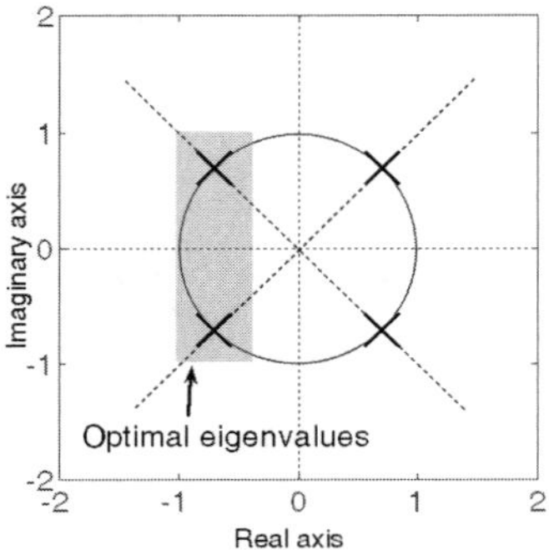

FIGURE 8.4
Solutions for equation (8.30) and the optimal eigenvalues ($r = 1$) for system $G(s) = \frac{1}{s^2}$.

We will see in Chapter 10 that systems satisfying $G(s) = -G^T(-s)$ are *loss-less* systems. For such systems, it can be proven (the theorem is often referred as Chebyshev theorem) that, given that $G(s) = \frac{b(s)}{a(s)}$ is loss-less, then the polynomial $a(s) + b(s)$ has negative real part roots. Therefore, for loss-less systems, the optimal eigenvalues with $r = 1$ are given by:

$$a(s) + b(s) = 0 \tag{8.34}$$

From the Letov theorem, some important considerations can be drawn in the extreme cases where $r \to 0$ or $r \to \infty$.

If r is very large (in the limit case $r \to \infty$), the term $\frac{1}{r}b(s)b(-s)$ in the Letov formula is negligible compared to term $a(s)a(-s)$. The optimal eigenvalues are given by $a(s)a(-s) = 0$. In this case, conclude that if the system has poles with negative real part, the optimal eigenvalues coincide with the open-loop system poles. If, instead, the system also has some poles with positive real part, then the optimal eigenvalues are given by the open-loop poles with negative real part and the sign reversed open-loop poles with positive real part. In conclusion, if the open-loop system is asymptotically stable then the optimal control with $r \to \infty$ is the one which leaves the system eigenvalues unchanged. The $r \to \infty$ case from a physical viewpoint really implies minimizing as much energy as possible at input (and at the limit therefore not acting on the system at all, given that by definition it is asymptotically stable).

When $r \to 0$, the term $a(s)a(-s)$ can be neglected with respect to $\frac{1}{r}b(s)b(-s)$ and the Letov formula can be approximated to $\frac{1}{r}b(s)b(-s) = 0$. This formula can calculate m optimal eigenvalues (where m is the order of polynomial $b(s)$). The other nm eigenvalues (n is the system order and so also the order of polynomial $a(s)$) can be obtained from the formula $s^{2(n-m)} = (-1)^{n-m+1}b_0^2r^{-1}$ where b_0 is the coefficient associated to the higher power in the polynomial $b(s)$, that is $b(s) = b_0 s^m + b_1 s^{m-1} + \ldots + b_m$. The eigenvalues are therefore arranged in a Butterworth configuration in a circle of radius $(\frac{b_0^2}{r})^{\frac{1}{2(n-m)}}$.

MATLAB® exercise 8.6

Let's reconsider the example in exercise 8.1 and calculate the optimal eigenvalues from the Letov formula for various values of r.

- If $r = 1$, then $\Delta(s) = (s^2-2s-3)(s^2+2-3s)+(s+2)(-s+2) = s^4-11s+13$. The solutions of $\Delta(s) = 0$ with negative real part are $\lambda_{1,ott} = -3.1070$ and $\lambda_{2,ott} = -1.1605$ and coincide with the optimal eigenvalues calculated in exercise 8.1.
- If $r = 10$, then the Letov formula produces $\Delta(s) \simeq (s^2 - 2s - 3)(s^2 + 2 - 3s)$ and so the optimal eigenvalues are given by the open-loop eigenvalues (if it is the case, sign changed) $\lambda_{1,ott} \simeq -1$ and $\lambda_{2,ott} \simeq -3$. To evaluate the accuracy of the approximation use MATLAB command:

  ```
  >> [K,P,E]=lqr(A,B,C'*C,10)
  ```

  $\lambda_{1,ott} = -1.0184$ are $\lambda_{2,ott} = -3.0104$ are obtained.
- If $r = 0.01$, then the Letov formula says that a closed-loop eigenvalue is given by the zero of the open-loop system, that is $\lambda_{1,ott} = -2$, whereas the other eigenvalue is given by the formula $s^2 = (-1)^2 \cdot 100$ (in fact $b_0 = 1$), and so $\lambda_{2,ott} = -10$.

  Let's consider the eigenvalues obtained by applying MATLAB® command

  ```
  >> [K,P,E]=lqr(A,B,C'*C,0.01)
  ```

  that is $\lambda_{1,ott} = -1.9629$ and $\lambda_{2,ott} = -10.3028$. Notice that even here the approximation from the Letov formula in the extreme case of very small r is more than sufficient.

Example 8.8

Given the system $G(s) = \frac{s+20}{(s^2-4s+8)^2}$ find the optimal eigenvalues with respect to r assuming that $J = \int_0^\infty (y^T y + u^T r u)dt$. Then fix r so that one optimal eigenvalue is equal to $\lambda = -5$.

Solution

System $G(s) = \frac{s+20}{(s^2-4s+8)^2}$ is a SISO system, in minimal form, and unstable with poles equal to $s_{1,2} = 2 \pm 2j$. Since the functional J is defined as per equation (8.24), all the assumptions that permit us to use the Letov formula to solve the assigned optimal control problem are respected.
So we have:

$$a(s)a(-s) + r^{-1}b(s)b(-s) = 0 \Rightarrow$$

$$(s^2 - 4s + 8)^2(s^2 + 4s + 8)^2 + r^{-1}(20 + s)(20 - s) = 0 \Rightarrow$$

$$s^4 - r^{-1}s^2 + 400r^{-1} + 64 = 0 \tag{8.35}$$

Solving the bi-quadratic equation and considering only the two solutions with negative real part, we have:

$$s_{ott,1,2} = -\sqrt{\frac{r^{-1} \pm \sqrt{r^{-2} - 1600r^{-1} - 256}}{2}}$$

To ensure that one of these solutions is equal to $\bar{\lambda}$, we have to impose that (8.35) equals zero for $s = \bar{\lambda}$:

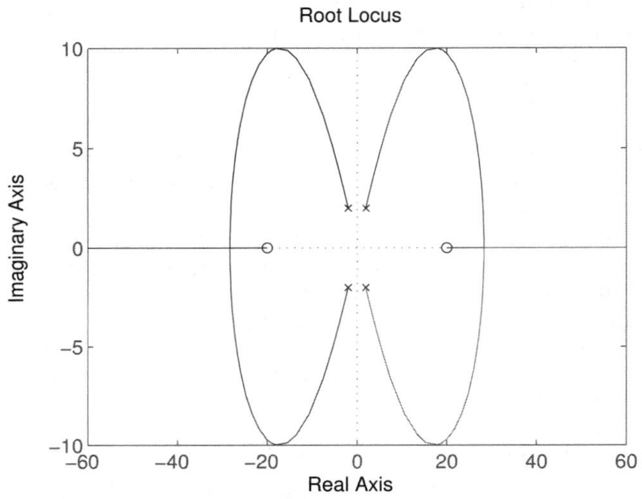

FIGURE 8.5
Root locus of $\tilde{G}(s) = \frac{400-s^2}{s^4+64}$.

$$\bar{\lambda}^4 - r^{-1}\bar{\lambda}^2 + 400r^{-1} + 64 = 0 \Rightarrow$$

$$r = \frac{\bar{\lambda}^2 - 400}{\bar{\lambda}^4 + 64}$$

Since $r > 0$, only $\bar{\lambda} < -20$ is admissible. So it is not possible to follow the assigned specifications.
The roots of the equation (8.35) can be studies by building an opportune root locus. Consider

$$s^4 - r^{-1}s^2 + 400r^{-1} + 64 = 0 \Rightarrow$$

$$1 + r^{-1}\frac{400 - s^2}{s^4 + 64} = 0$$

By studying the root locus of the fictitious transfer function $\tilde{G}(s) = \frac{400-s^2}{s^4+64}$, all the solutions of the equation (8.35) can be obtained. To know the optimal eigenvalues with respect to r, we have to refer to the left half plane of this root locus. Notice there are no values of r^{-1} (and so of r) which ensure real solutions with $-20 < \bar{\lambda} < 0$.

8.6 Linear quadratic regulator with cross-weighted cost

There exist several variations of the LQR problem presented above. An important variation derives from the observation that the $\mathbf{z}(t)$ variables to keep small may contain a feedthrough term. In fact, if we consider $\mathbf{z}(t) = \mathrm{C}\mathbf{x}(t) + \mathrm{D}\mathbf{u}(t)$, the index to minimize becomes:

$$J = \int_0^\infty (\mathbf{x}^T \mathrm{C}^T \mathrm{C}\mathbf{x} + 2\mathbf{x}^T \mathrm{C}^T \mathrm{D}\mathbf{u} + \mathbf{u}^T (\mathrm{R} + \mathrm{D}^T \mathrm{D})\mathbf{u})dt \tag{8.36}$$

For this reason, in LQR problems the more general definition of the index to minimize

$$J = \int_0^\infty (\mathbf{x}^T \mathrm{Q}\mathbf{x} + 2\mathbf{x}^T \mathrm{N}\mathbf{u} + \mathbf{u}^T \mathrm{R}\mathbf{u})dt \tag{8.37}$$

may be considered.

Index (8.37) appears when the variables to keep small contain derivatives of the state variables, but cross-coupled costs also appear when frequency-dependent weighting terms are used.

In this case, the Riccati equation becomes:

$$\mathrm{P}(\mathrm{A}-\mathrm{B}\mathrm{R}^{-1}\mathrm{N}^T)+(\mathrm{A}-\mathrm{B}\mathrm{R}^{-1}\mathrm{N}^T)^T\mathrm{P}-\mathrm{P}\mathrm{B}\mathrm{R}^{-1}\mathrm{B}^T\mathrm{P}+(\mathrm{Q}-\mathrm{N}\mathrm{R}^{-1}\mathrm{N}^T) = 0 \tag{8.38}$$

with $\mathrm{P} \in \mathbb{R}^{n \times n}$. If $\overline{\mathrm{P}}$ is the solution, the optimum gains are now given by:

$$\mathrm{K}_{ott} = \mathrm{R}^{-1}(\mathrm{N}^T + \mathrm{B}^T\overline{\mathrm{P}}) \tag{8.39}$$

The LQR problem defined by index (8.37) may be also solved in MATLAB® with command `lqr`.

8.7 Finite-horizon linear quadratic regulator

For system (8.1) let us consider now the following performance index:

$$J = \int_t^T (\mathbf{x}^T(\tau)\mathrm{Q}\mathbf{x}(\tau) + \mathbf{u}^T(\tau)\mathrm{R}\mathbf{u}(\tau))d\tau \tag{8.40}$$

which assumes that the upper bound of the integral in the performance index is a finite time T rather than infinite and that the lower bound is T rather than zero.

The gains of the optimal control law are now given by:

$$\mathrm{K}_{ott} = \mathrm{R}^{-1}\mathrm{B}^T\mathrm{P(t)} \tag{8.41}$$

where P(t) is now a time-varying matrix given by the solution of the *differential Riccati equation*:

$$\dot{\mathrm{P}} + \mathrm{Q} - \mathrm{PBR}^{-1}\mathrm{B}^T\mathrm{P} + \mathrm{PA} + \mathrm{A}^T\mathrm{P} = 0 \tag{8.42}$$

with $\mathrm{P}(T) = 0$.

Equation (8.42) is difficult to analytically solve for systems with order higher than two. It is usually numerically solved with an iterative backward in time procedure.

The solution of the infinite horizon problem is recovered from that of equation (8.42) by taking:

$$\overline{\mathrm{P}} = \lim_{t\to-\infty} \mathrm{P}(t,T) = \lim_{T\to\infty} \mathrm{P}(0,T) \tag{8.43}$$

In the following example the use of MATLAB® symbolic computation to solve a simple differential Riccati equation is illustrated.

MATLAB® exercise 8.7

Consider system:

$$\dot{x} = x + u \tag{8.44}$$

and the performance index (8.40) with $Q = 1$ and $R = 1$. Since a state-space realization of system (8.44) is given by $A = 1$ and $B = 1$, equation (8.42) becomes:

$$\dot{P} - P^2 + 2P + 1 = 0 \tag{8.45}$$

It can be solved with the MATLAB command:

```
>> dsolve('DP=P^2-2*P-1')
```

obtaining

$$P(t) = 1 - \sqrt{2}\tanh(\sqrt{2}t + \sqrt{2}c_1) \tag{8.46}$$

where the constant c_1 is obtained by imposing $P(T) = 0$.
The solution of the infinite horizon problem is obtained as:

$$\overline{\mathrm{P}} = \lim_{t\to-\infty} \mathrm{P}(t,T) = 1 + \sqrt{(2)} \tag{8.47}$$

that is the same result that can be obtained by solving the algebraic Riccati equation (8.4) with command:

```
>> are(1,1,1)
```

8.8 Optimal control for discrete-time linear systems

To conclude this chapter, let's look briefly at discrete-time systems. Consider a discrete-time system in state-space form:

$$\begin{aligned} \mathbf{x}_{k+1} &= \mathrm{A}\mathbf{x}_k + \mathrm{B}\mathbf{u}_k \\ \mathbf{y}_k &= \mathrm{C}\mathbf{x}_k \end{aligned} \tag{8.48}$$

The quadratic index in this case is defined as:

$$J = \sum_{k=0}^{\infty} (\mathbf{x}_k^T \mathrm{Q} \mathbf{x}_k + \mathbf{u}_k^T \mathrm{R} \mathbf{u}_k) dt \tag{8.49}$$

with Q positive semi-definite and R positive definite. The gain matrix of control law $\mathbf{u}_k = -\mathrm{K}\mathbf{x}_k$ which minimizes index (8.49) is found by resolving algebraic Riccati equation for discrete-time systems:

$$\mathrm{A}^T\mathrm{P}\mathrm{A} - \mathrm{P} - \mathrm{A}^T\mathrm{P}\mathrm{B}(\mathrm{R} + \mathrm{B}^T\mathrm{P}\mathrm{B})^{-1}\mathrm{B}^T\mathrm{P}\mathrm{A} + \mathrm{Q} = 0 \tag{8.50}$$

and by making

$$\mathrm{K} = (\mathrm{B}^T\mathrm{P}\mathrm{B} + \mathrm{R})^{-1}\mathrm{B}^T\mathrm{P}\mathrm{A} \tag{8.51}$$

MATLAB® exercise 8.8

Consider the discrete-time LTI MIMO system with state-space matrices:

$$\mathrm{A} = \begin{bmatrix} -2 & 0 & 0 \\ 0 & 1 & 0 \\ 0 & 0 & 0.2 \end{bmatrix}; \mathrm{B} = \begin{bmatrix} 1 & 0.1 \\ 2 & 5 \\ 0.3 & 3 \end{bmatrix}; \mathrm{C} = \begin{bmatrix} 1 & 0 & 1 \\ -1 & 2 & 1 \end{bmatrix} \tag{8.52}$$

and the optimal control problem (8.49) with $\mathrm{Q} = \mathrm{C}^T\mathrm{C}$ and $\mathrm{R} = 2\mathrm{I}$.
Define system (8.52) in MATLAB

```
>> A=[-2 0 0; 0 1 0; 0 0 0.2]
>> B=[1 0.1; 2 5; 0.3 3]
>> C=[1 0 1; -1 2 1]
>> D=zeros(2)
```

and then use the command `dlqr` to calculate the linear quadratic regulator for discrete-time LTI systems:

```
>> [K,P,E]=dlqr(A,B,C'*C,2*eye(2))
```

One obtains the closed-loop optimal eigenvalues: $\lambda_1 = -0.3530$, $\lambda_2 = 0.3753$, and $\lambda = 0.0050$ as it can be verified with command:

```
>> eig(A-B*K)
```

8.9 Exercises

1. Given the system $G(s) = \frac{s+2}{s^2+4s+6}$ calculate the optimal controller that minimizes the index defined by $\mathrm{Q} = \mathrm{C}^T\mathrm{C}$ in these three cases: $r = r_1 = 1$, $r = r_2 = 0.01$ and $r = r_3 = 20$. Verify the performance of the control law obtained.

2. Given the system with transfer function $G(s) = \frac{2s-1}{s(s-1)}$ calculate the optimal eigenvalues with respect to r, if the index to optimize is $J = \int_0^\infty (\mathbf{y}^T\mathbf{y} + \mathbf{u}^T r\mathbf{u})dt$. Then calculate the characteristic values of the system fixing r so that the optimal eigenvalue is $\lambda = -\sqrt{2}$.
3. Design the linear quadratic regulator ($r = 2$) for the system $G(s) = \frac{s^2+2s+1}{s^3-s^2+5s+3}$ and calculate the optimal eigenvalues.
4. Given the system with state-space matrices:

$$\mathrm{A} = \begin{bmatrix} 0 & 1 \\ 3 & -2 \end{bmatrix}; \mathrm{B} = \begin{bmatrix} 0 \\ 1 \end{bmatrix}; \mathrm{C} = \begin{bmatrix} -1 & 1 \end{bmatrix}$$

design the linear quadratic regulator with $\mathrm{Q} = \mathrm{C}^T\mathrm{C}$ and $r = 5$.
5. Given the system with state-space matrices:

$$\mathrm{A} = \begin{bmatrix} 0 & 1 & 0 \\ -1 & 0 & 0 \\ 0 & 0 & 3 \end{bmatrix}; \mathrm{B} = \begin{bmatrix} 0 \\ 1 \\ -1 \end{bmatrix}; \mathrm{C} = \begin{bmatrix} 1 & 1 & 1 \end{bmatrix}$$

design the linear quadratic regulator with $\mathrm{Q} = \mathrm{I}$ and $r = 7$.

9

Closed-loop balanced realization

CONTENTS

In this chapter we will look at synthesizing a compensator for a high order system. One strategy is to consider a lower order model of the process, $G_r(s)$, and design a lower order compensator $C_r(s)$. In this way, the regulator is bound to work well for the lower order system (see Figure 9.1), but not necessarily for the original model $G(s)$ (see Figure 9.2). Therefore, the efficacy of the project needs to be verified (with numerical simulations).

Another strategy is to design $C(s)$ by assigning poles (design of the observer and of the control law) on the basis of system $G(s)$. In this way, an n order compensator is obtained which can be approximated through model order reduction. This case is *direct approximation of the compensator*, whereas designing a lower order compensator is known as indirect approximation of the compensator.

In both techniques, approximating the compensator is quite separate from designing it. Below, we'll see how closed-loop balancing is performed by using the CARE equation introduced in the previous chapter and a dual equation, the FARE equation (solving the dual problem of optimal filtering). With this technique, instead of approximating the controller or the process prior to or after the synthesis, the controller and process are approximated simultaneously and during synthesis.

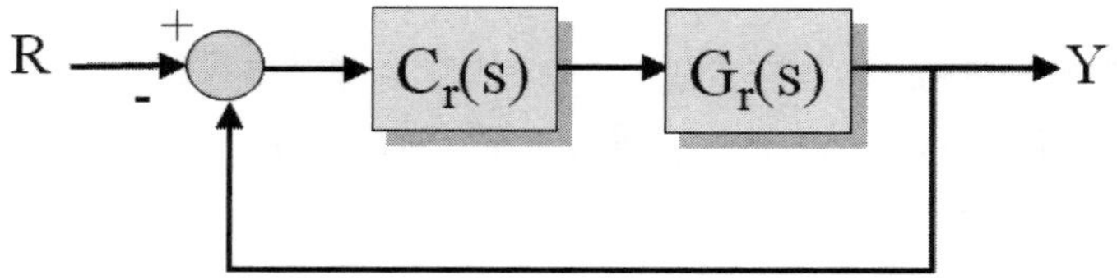

FIGURE 9.1
Control scheme with lower order compensator applied to a lower order system $G_r(s)$.

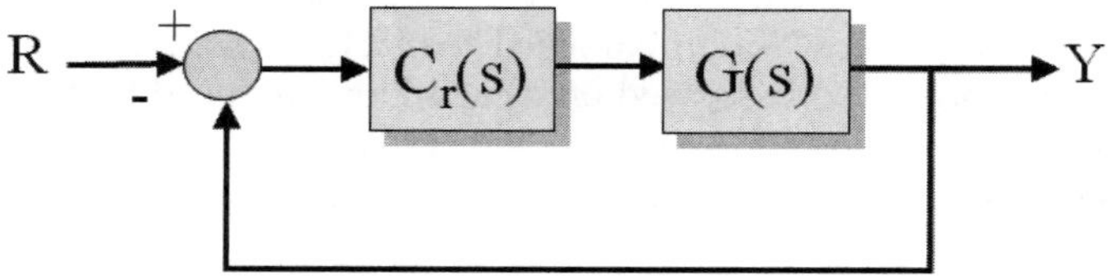

FIGURE 9.2
Control scheme with lower order compensator applied to a $G(s)$ system.

9.1 Filtering Algebraic Riccati Equation

The dual CARE equation takes its name from FARE (Filtering Algebraic Riccati Equation) and is defined as:

$$\mathrm{A}\Pi + \Pi\mathrm{A}^T - \Pi\mathrm{C}^T\mathrm{C}\Pi + \mathrm{B}\mathrm{B}^T = 0 \tag{9.1}$$

Just as CARE resolves optimal control problem once a certain functional needs minimizing, FARE can design an optimal observer according to a particular criteria.

Referring to Figure 9.3, remember that optimal control of a linear regulator requires knowing the entire state vector $\mathbf{x}(t)$ or should all the state variables not be measurable, an estimate $\tilde{\mathbf{x}}(t)$ of the state via an observer. To design an optimal controller, even designing the observer must be based on optimizing a particular criteria. Designing a controller means finding gains K and h on the basis of established criteria. In the previous chapter an optimal strategy was found based on minimizing a quadratic functional in choosing the K gains. An optimization criteria can be dually defined even for an observer project.

Remember that the prime condition which must be satisfied in an observer project is to make matrix A_o a stability matrix. Once the error is defined as $\mathbf{e}(t) = \mathbf{x}(t) - \tilde{\mathbf{x}}(t)$, then the error system dynamic is given by:

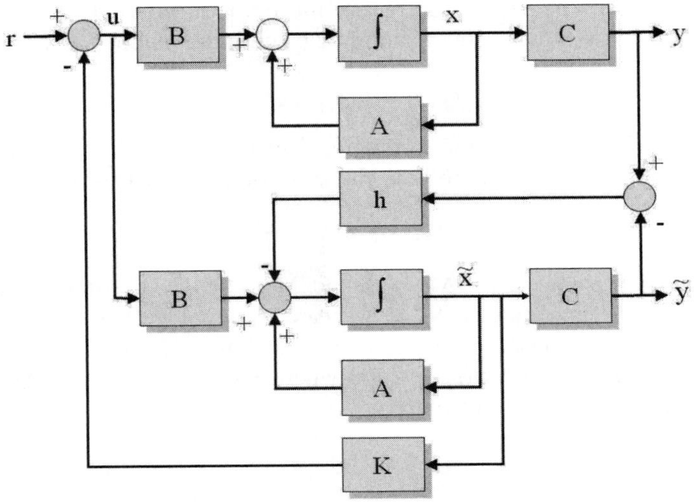

FIGURE 9.3
Control scheme with a state linear regulator and an asymptotic observer.

$$\dot{\mathbf{e}}(t) = \mathrm{A}_0\mathbf{e}(t) \tag{9.2}$$

so that, to make $\mathbf{x}(t) \simeq \tilde{\mathbf{x}}(t)$, that is that $\mathbf{e}(t) = e^{\mathrm{A}_0 t}\mathbf{e}(0) \to 0$, the h gains must be chosen such that all the eigenvalues of $\mathrm{A}_0 = \mathrm{A} - \mathrm{hC}$ have negative real part.

Generally, noise added to the state variables, that is an uncertainty which makes the variables of state differ from the mathematical model, must also be considered. Moreover, a measurement noise must also be considered. All this leads to formulating the following state-space model:

$$\begin{aligned} \dot{\mathbf{x}}(t) &= \mathrm{A}\mathbf{x}(t) + \mathrm{B}\mathbf{u}(t) + \mathbf{d}(t) \\ \mathbf{y}(t) &= \mathrm{C}\mathbf{x}(t) + \mathbf{v}(t) \end{aligned} \tag{9.3}$$

where $\mathbf{d}(t)$ (noise on the state variables) and $\mathbf{v}(t)$ (measurement noise) are stochastic signals. Their value is not known exactly but their statistics are. Suppose that $\mathbf{d}(t)$ and $\mathbf{v}(t)$ are random signals with zero mean and suppose we know the covariance matrices which characterize them:

$$\mathrm{M}_d = E\{\mathbf{d}(t)\mathbf{d}(t)^T\}$$

$$\mathrm{M}_v = E\{\mathbf{v}(t)\mathbf{v}(t)^T\}$$

where $\mathrm{M}_d \in \mathbb{R}^{n\times n}$ and $\mathrm{M}_v \in \mathbb{R}^{p\times p}$. $E\{\cdot(x(t))\} = \int_{-\infty}^{\infty} x p_{x(t)}(x)dt$ is the expected value of the stochastic signal $x(t)$ with a probability density function of $p_{x(t)}(x)$. Even the error system at this point is non-deterministic. Let's establish the eigenvalues so as to minimize the index defined by:

$$\bar{J} = E\{(\mathbf{x}(t) - \tilde{\mathbf{x}}(t))^T(\mathbf{x}(t) - \tilde{\mathbf{x}}(t))\} \tag{9.4}$$

Optimal filtering is the dual problem of optimal control. Optimal gains are given by $\mathrm{h}_{ott}^T = \mathrm{M}_v^{-1}\mathrm{C}\Pi$ where Π is the solution of the Riccati equation:

$$\mathrm{A}\Pi + \Pi\mathrm{A}^T - \Pi\mathrm{C}^T\mathrm{M}_v^{-1}\mathrm{C}\Pi + \mathrm{M}_d = 0 \tag{9.5}$$

M_v is a $p \times p$ matrix. If $\mathrm{M}_v = \mathrm{I}$ and $\mathrm{M}_d = \mathrm{B}\mathrm{B}^T$ in equation (9.5), the result is the FARE equation (9.1). As regards signal $\mathbf{v}(t)$, this choice means that its covariance matrix is unitary. As for signal $\mathbf{d}(t)$ making $\mathrm{M}_d = \mathrm{B}\mathrm{B}^T$ means hypothesizing that $\mathbf{d}(t) = \mathrm{B}\xi(t)$ where $\xi(t)$ is a stochastic signal which also has a unitary covariance matrix.

9.2 Computing the closed-loop balanced realization

The CARE (8.18) and FARE (9.1) equations are normalized nonlinear given that $\mathrm{R} = \mathrm{I}$ and $\mathrm{M}_v = \mathrm{I}$. With these two equations and using the same procedure which produces open-loop balanced forms from gramians equations, we can obtain a closed-loop balanced form. To simplify, we'll refer to the closed-loop balanced realization form based on these normalized equations even though the procedure is very general and can be applied to the corresponding non-normalized equations.

Let's consider the CARE and FARE equations:

$$\mathrm{A}^T\mathrm{P} + \mathrm{P}\mathrm{A} - \mathrm{P}\mathrm{B}\mathrm{B}^T\mathrm{P} + \mathrm{C}^T\mathrm{C} = 0 \tag{9.6}$$

$$\mathrm{A}\Pi + \Pi\mathrm{A}^T - \Pi\mathrm{C}^T\mathrm{C}\Pi + \mathrm{B}\mathrm{B}^T = 0 \tag{9.7}$$

and let's assign $\bar{\mathrm{P}}$ and $\bar{\Pi}$ as the only definite positive solutions of CARE and FARE. These two matrices depend on the reference system in the same way the gramians depend on the reference system (this can be proved by following the same steps done for the case of open-loop balanced realization). The $\bar{\Pi}\bar{\mathrm{P}}$ eigenvalues however do not depend on the reference system. They allow to define system invariants, called *characteristic values of the system.*

Definition 14 (Characteristic values of the system) *The characteristic values of a system in minimal form $\mu_1 \geq \mu_2 \geq \ldots \geq \mu_n$ are the square roots in descending order of the $\bar{\Pi}\bar{\mathrm{P}}$ eigenvalues. They are invariant quantities, i.e., they do not depend on the state-space representation of the system.*

Having defined the characteristic values of a system, the issue of closed-loop balanced realization lies in finding the form for which the CARE and FARE equations produce the same diagonal solution, i.e., $\bar{\Pi} = \bar{P} = \text{diag}(\mu_1, \mu_2, \ldots, \mu_n)$.

Definition 15 (Closed-loop balanced realization) *Given a linear time-invariant system in minimal form, the closed-loop balanced realization is that realization in which* $\bar{\Pi} = \bar{P} = \text{diag}(\mu_1, \mu_2, \ldots, \mu_n)$*, where* $\bar{P}$ *and* $\bar{\Pi}$ *are the only positive definite solutions for CARE and FARE.*

Note that the hypothesis which makes it possible to define the closed-loop balanced realization of a system is the existence of the CARE and FARE solutions, that is that the system is completely controllable and observable. As opposed to the case for open-loop balancing, it is no longer necessary that the open-loop system is asymptotically stable. Besides, since the system is controllable and optimal control is being designed, the closed-loop system is guaranteed to be asymptotically stable.

This balancing technique arose from the need to balance systems with little damping. Known as large space flexible structures they are used to model antennae or light satellites with solar energy panels. In particular, often distributed parameter models for these structures or concentrated parameter models with a large number of state variables are formulated. The closed-loop balancing of this class of structures arose from the need to design controllers which could stabilize the oscillations of these very flexible structures. Open-loop balancing for these types of structures is difficult numerically because they have little damping. Closed-loop balancing has the advantage of not causing numerical errors in finding the solution and it allows the controller to be designed while effecting the balancing.

Closed-loop balancing facilitates designing the optimal regulator and observer at the same time by proceeding analogously to the open-loop case to reduce system order. In the next paragraph, we will analyze in detail the issues surrounding lower order models based on closed-loop equilibrium.

9.3 Procedure for closed-loop balanced realization

The procedure for closed-loop balanced realization is analogous to the one for open-loop, on condition that the CARE and FARE equations are used instead of the two Lyapunov equations of the two gramians. The procedure is illustrated below. Note that the FARE solution is calculated first, then singular value decomposition is applied to matrix $\bar{\Pi}$ and from this the transformation matrix P_1 is defined. In the new reference system (where it's easy to verify $\tilde{\Pi} = I$) the CARE solution then needs calculating, then proceeding to its singular value decomposition and then defining the transformation matrix T_2.

MATLAB® exercise 9.1

The procedure is reported below for closed-loop balanced realization.

```
function [system_bal,G,PIGREEK_bal,P_care_bal]=balrealmcc(system)
% closed-loop balanced realization
%    [system_bal, G, PIGREEK_bal, P_care_bal]=balrealmcc(system)
% system is defined in state-space form
A=system.A;
B=system.B;
C=system.C;
%rank(ctrb(system))
%rank(obsv(system))
%controllability and observability hypotheses
PIGREEK=are(A',C'*C,B*B');  %FARE
[Uc,Sc2,Vc]=svd(PIGREEK);
P1=Vc*sqrt(Sc2);
Atilde=inv(P1)*A*P1;
Btilde=inv(P1)*B;
Ctilde=C*P1;
%%%verify: Wtildec2=I
PIGREEKtilde=are(Atilde',Ctilde'*Ctilde,Btilde*Btilde'); %FARE
Pcaretilde=are(Atilde,Btilde*Btilde',Ctilde'*Ctilde);  %CARE
[Uo,So2,Vo]=svd(Pcaretilde);
T2=Vo*(So2)^(-1/4);
Abil=inv(T2)*Atilde*T2;
Bbil=inv(T2)*Btilde;
Cbil=Ctilde*T2;
system_bal=ss(Abil,Bbil,Cbil,0);
PIGREEK_bal=are(Abil',Cbil'*Cbil,Bbil*Bbil'); %New FARE
P_care_bal=are(Abil,Bbil*Bbil',Cbil'*Cbil);  %CARE
%%%%%%%%%%%%%Verify: calculate the characteristic values of system
Pcare=are(A,B*B',C'*C);  %CARE
G=sqrt(eig(PIGREEK*Pcare));
```

This function can be applied to a system in state-space form. For instance, given the system:

```
>> A=[-0.2 0.5 0 0 0;
0 -0.5 1.6 0 0;
0 0 -14.3 87.7 0;
0 0 0 -25 75;
0 0 0 0 -10]
>> B=[0 0 0 0 30]'
>> C=[1 0 0 0 0]
>> system=ss(A,B,C,0)
```

the closed-loop balanced realization is constructed with commands:

```
>> [system_bal,G,PIGREEK_bal,P_care_bal]=balrealmcc(system)
```

We get the characteristic values $\mu_1 = 3.9859$, $\mu_2 = 0.7727$, $\mu_3 = 0.1276$, $\mu_4 = 0.0101$ and $\mu_5 = 0.0003$, suggesting the choice of $r = 3$ for the order of the reduced order model.

9.4 Reduced order models based on closed-loop balanced realization

$\mathbf{X}^*$ represents the reference system for the closed-loop balanced form. In this reference system the CARE solution, $\bar{\mathrm{P}}^*$, is diagonal and the optimal gains are of the form:

$$K^*_{ott} = \mathrm{B}^{*T}\mathrm{diag}(\mu_1, \mu_2, \ldots, \mu_n)$$

Furthermore, $\mathrm{A}^*_c = \mathrm{A}^* - \mathrm{B}^*\mathrm{K}^*_{ott} = \mathrm{A}^* - \mathrm{B}^*\mathrm{B}^{*T}\bar{\mathrm{P}}^*$ certainly has all eigenvalues with negative real part.

If index r is such that the system characteristic values can be divided into two groups so that $\mu_1 \geq \mu_2 \geq \ldots \geq \mu_r \gg \mu_{r+1} \geq \ldots \geq \mu_n$, then a lower order model could be constructed which only accounted for the first r state variables of the system. In other words, instead of feeding back the n state variables, only r are fed back.

So, in matrix A^*_c, we will take into account matrix $\bar{\mathrm{P}}^*$ defined by

$$\bar{\mathrm{P}}^* = \begin{bmatrix} \mu_1 & & & & & \\ & \ddots & & & & \\ & & \mu_r & & & \\ & & & 0 & & \\ & & & & \ddots & \\ & & & & & 0 \end{bmatrix}$$

Corresponding to this approximation, the control law becomes:

$$\mathbf{u} = -\mathrm{K}^*_{ott}\mathbf{x}^* = -\mathrm{B}^{*T}\bar{\mathrm{P}}^*\mathbf{x}^* = -\mathrm{B}^{*T} \begin{bmatrix} \mu_1 x^*_1 \\ \vdots \\ \mu_r x^*_r \\ 0 \\ \vdots \\ 0 \end{bmatrix}$$

If the discarded characteristic values $\mu_{r+1}, \ldots, \mu_n$ are effectively small, the signals $\mu_{r+1}x^*_{r+1}, \ldots, \mu_n x^*_n$ contribute minimally to the feedback and so it should be expected that matrix A^*_c remains a stability matrix. In other words, the characteristic values $\mu_1, \ldots, \mu_n$ have the weight to take into account how important the corresponding state variables are in the feedback. When there are two groups of characteristic values of very differing orders ($\mu_r \gg \mu_{r+1}$), some system variables have less effect in the feedback and the closed-loop system remains stable even using the first r state variables in the feedback. Obviously, in this case, the closed-loop system is stable, but control system performance could fall off. Alternatively, what happens when an open-loop optimum approximation is made is that the closed-loop system destabilizes because an important state variable was discarded to carry out the feedback (i.e., associated with a high characteristic value). It is not an absolute given that the strongly controllable and observable parts are the most important for guaranteeing closed-loop stability.

Suppose the system is closed-loop balanced. It has been seen that if $\mu_r \gg \mu_{r+1}$, then the state vector can be partitioned: $\mathbf{x} = \begin{bmatrix} \mathbf{x}_1 \\ \mathbf{x}_2 \end{bmatrix}$, where $\mathbf{x}_1$ represents the first r components and $\mathbf{x}_2$ the remaining $n-r$ components. Correspondingly, matrices A, B and C can be partitioned as:

$$\mathrm{A} = \begin{bmatrix} \mathrm{A}_{11} & \mathrm{A}_{12} \\ \mathrm{A}_{21} & \mathrm{A}_{22} \end{bmatrix}; \mathrm{B} = \begin{bmatrix} \mathrm{B}_1 \\ \mathrm{B}_2 \end{bmatrix}; \mathrm{C} = \begin{bmatrix} \mathrm{C}_1 & \mathrm{C}_2 \end{bmatrix}$$

Remembering that $\mathrm{K}_{ott} = \mathrm{B}^T\bar{\mathrm{P}}$ with $\bar{\mathrm{P}} = \begin{bmatrix} \mu_1 & & & \\ & \mu_2 & & \\ & & \ddots & \\ & & & \mu_n \end{bmatrix} = \begin{bmatrix} \bar{\mathrm{P}}_1 & \\ & \bar{\mathrm{P}}_2 \end{bmatrix}$, the closed-loop matrix is given by:

$$\mathrm{A}_c = \mathrm{A} - \mathrm{BK}_{ott} = \mathrm{A} = \begin{bmatrix} \mathrm{A}_{11} & \mathrm{A}_{12} \\ \mathrm{A}_{21} & \mathrm{A}_{22} \end{bmatrix} - \begin{bmatrix} \mathrm{B}_1 \\ \mathrm{B}_2 \end{bmatrix} \begin{bmatrix} \mathrm{B}_1^T & \mathrm{B}_2^T \end{bmatrix} \begin{bmatrix} \bar{\mathrm{P}}_1 & 0 \\ 0 & \bar{\mathrm{P}}_2 \end{bmatrix} =$$

$$= \begin{bmatrix} \mathrm{A}_{11} - \mathrm{B}_1\mathrm{B}_1^T\bar{\mathrm{P}}_1 & \mathrm{A}_{12} - \mathrm{B}_1\mathrm{B}_2^T\bar{\mathrm{P}}_2 \\ \mathrm{A}_{21} - \mathrm{B}_2\mathrm{B}_1^T\bar{\mathrm{P}}_1 & \mathrm{A}_{22} - \mathrm{B}_2\mathrm{B}_2^T\bar{\mathrm{P}}_2 \end{bmatrix}$$

We are certain that the eigenvalues of those matrices have negative real part. Whereas, let's consider the lower order model (i.e., neglecting $n-r$ state variables, or rather if $\bar{\mathrm{P}}_2 = 0$), the matrix becomes:

$$\tilde{\mathrm{A}}_c = \begin{bmatrix} \mathrm{A}_{11} - \mathrm{B}_1\mathrm{B}_1^T\bar{\mathrm{P}}_1 & \mathrm{A}_{12} \\ \mathrm{A}_{21} - \mathrm{B}_2\mathrm{B}_1^T\bar{\mathrm{P}}_1 & \mathrm{A}_{22} \end{bmatrix}$$

Now, the state linear regulator is applied on r state variables instead of on all the n state variables and the closed-loop system may also be unstable as shown in the next example.

MATLAB® exercise 9.2

Consider the continuous-time LTI system with state-space matrices:

$$\mathrm{A} = \begin{bmatrix} 1 & 0 & 0 & 0 \\ 0 & 10 & 0 & 0 \\ 0 & 0 & 2 & 0 \\ 0 & 0 & 0 & 3 \end{bmatrix}; \mathrm{B} = \begin{bmatrix} 1 \\ 1 \\ 1 \\ 1 \end{bmatrix}; \mathrm{C} = \begin{bmatrix} 1 & 1 & 1 & 1 \end{bmatrix} \tag{9.8}$$

Define the system in MATLAB®

```
>> A=diag([1 10 2 3])
```

```
>> B=ones(4,1)
>> C=ones(1,4)
>> system=ss(A,B,C,0)
```

and compute the closed-loop balanced realization with the procedure described in Section 9.3:

```
>> [systemb, G, Pgreekb, Pcareb]=balrealmcc(system)
```

Now, let's consider $r = 2$ and impose $\bar{\mathrm{P}}_2$

```
>> Pcarebred=Pcareb
>> Pcarebred(3:4,3:4)=zeros(2)
```

We calculate the gains

```
>> Kred=systemb.b'*Pcarebred
```

and the closed-loop eigenvalues

```
>> eig(systemb.a-systemb.b*Kred)
```

One obtains $\lambda_{1,2} = 0.0655 \pm 14.6635j$, $\lambda_3 = -0.7484$ and $\lambda_4 = 0.2539$ which show that the closed-loop system is not stable. Instead, if the full control is applied:

```
>> K=systemb.b'*Pcareb
>> eig(systemb.a-systemb.b*K)
```

The closed-loop eigenvalues are: $\lambda_1 = -10.3287$, $\lambda_2 = -3.9679$, $\lambda_3 = -1.3195$, $\lambda_4 = -2.4151$ and the closed-loop system is stable.

It has already been said that, if $\bar{\mathrm{P}}_2$ is made up of small coefficients, the approximation deriving from the fact that we are applying the linear regulator only on r state variables instead of on all the n state variables, is allowed. Let's now find out to what point the approximation is allowed or in other words to what point is $\tilde{\mathrm{A}}_c$ a stability matrix. To do this, the contribution $\bar{\mathrm{P}}_2$ can be interpreted as a perturbation term

$$\tilde{\mathrm{A}}_c = \mathrm{A}_c + \Delta\mathrm{A}_c$$

with $\Delta\mathrm{A}_c = \begin{bmatrix} 0 & +\mathrm{B}_1\mathrm{B}_2^T\bar{\mathrm{P}}_2 \\ 0 & +\mathrm{B}_2\mathrm{B}_2^T\bar{\mathrm{P}}_2 \end{bmatrix}$.

Generally, given matrices

$$\tilde{\mathrm{A}} = \mathrm{A} + \Delta\mathrm{A}$$

let's consider what specifics $\Delta\mathrm{A}$ must possess as long as, given that A is a stability matrix, also $\tilde{\mathrm{A}}$ is a stability matrix.

In certain particular cases the problem can be simply resolved. Think of the case in which matrix A is an inferior triangular matrix. Any type of perturbation which modifies the A coefficients below the diagonal means that $\tilde{\mathrm{A}}$ is a stability matrix. So, if matrix $\Delta\mathrm{A}$ has a particular structure, it does not affect the stability of $\tilde{\mathrm{A}}$. Therefore, the conditions depend on the fact that the perturbation is localized onto some terms or spread across all the terms of the matrix.

More generally, the only information we have about matrix A, is that it is a stability matrix. The conditions on $\Delta\mathrm{A}$ are of the type

$$\|\Delta\mathrm{A}\| < \beta \tag{9.9}$$

so that if the 2-norm of this matrix is less than β, the perturbed system is bound to remain stable. There are various methods of calculating β. The least conservative (to which the highest value of β is associated) establishes that:

$$\beta_{max} = \frac{1}{\|(s\mathrm{I} - \mathrm{A})^{-1}\|_\infty} \tag{9.10}$$

where $\|(s\mathrm{I} - \mathrm{A})^{-1}\|_\infty$ is the H_∞ norm of matrix $(s\mathrm{I} - \mathrm{A})^{-1}$:

$$\|(s\mathrm{I} - \mathrm{A})^{-1}\|_\infty = \sup_{\omega}\{\|(j\omega\mathrm{I} - \mathrm{A})^{-1}\|_S : \omega \in \mathbb{R}\} \tag{9.11}$$

Matrix $(s\mathrm{I} - \mathrm{A})$ is a complex matrix function of ω. As ω varies, the matrix whose spectral norm is greatest should be considered. β_{max} is the greatest value which assures that if relation (9.9) is verified, matrix $\tilde{\mathrm{A}}$ is still stable.

Returning to the more specific case under examination, note that matrix $\Delta\mathrm{A}$ can be written thus:

$$\Delta\mathrm{A} = \tilde{\mathrm{B}}\tilde{\mathrm{B}}^T\Delta\mathrm{P} \tag{9.12}$$

with $\Delta\mathrm{P} = \begin{bmatrix} 0 & 0 \\ 0 & \mathrm{P}_2 \end{bmatrix}$.

By using consistent norms:

$$\|\Delta\mathrm{A}\| = \|\tilde{\mathrm{B}}\tilde{\mathrm{B}}^T\Delta\mathrm{P}\| \leq \|\tilde{\mathrm{B}}\tilde{\mathrm{B}}^T\| \cdot \|\Delta\mathrm{P}\| \leq \|\tilde{\mathrm{B}}\tilde{\mathrm{B}}^T\| \cdot \mu_{r+1} \tag{9.13}$$

Since P_2 is a diagonal matrix, its norm will be less than μ_{r+1}. In this way, a relationship between the uncertainty matrix and the characteristic value μ_{r+1} (which corresponds to the first state variable neglected in the closed-loop reduced model) is established. If $f(\mathrm{A}_c) = \beta$, then

$$\mu_{r+1} \leq \frac{f(\mathrm{A}_c)}{\|\tilde{\mathrm{B}}\tilde{\mathrm{B}}^T\|} \tag{9.14}$$

μ_{r+1} is a quantity which depends on A_c and allows us to establish the condition in which $\tilde{\mathrm{A}}_c$ is a matrix of (asymptotic) stability.

From the closed-loop balanced realization, the characteristic values can therefore be calculated as well as that which satisfies (9.14). At this point (having found index r), the number of state variables which must be used in feedback is established. Naturally, performance has fallen off as you may note from the performance index $J = \mathbf{x}^T(0)\mathrm{P}\mathbf{x}(0)$ which shows an error proportional to neglected coefficients $\mu_{r+1}, \dots, \mu_n$ (in fact P does not contain sub-matrix P_2).

The direct truncation and singular perturbation techniques for model order reduction can be applied to closed-loop balanced realization exactly like in the case of open-loop balanced realization widely discussed in Chapter 6. We mention here another result on reduced order models based on the closed-loop balanced realization which is important for the design of low-order compensators. Starting from a system in closed-loop balanced realization, the reduced order model obtained with direct truncation is still in a closed-loop balanced form. When A_{22} is stable, the singular perturbation approximation can be applied and the reduced order model is also closed-loop balanced.

9.5 Closed-loop balanced realization for symmetrical systems

As we saw in Chapter 7 in the case of open-loop balanced realization, symmetrical systems have the advantage of certain properties which simplify calculating the balancing. Very similar properties are also valid for closed-loop balanced realization.

Let's consider the CARE equation (8.18) and apply the relations characteristic of symmetrical systems ($C = B^T T$, $B = T^{-1}C^T$, $A = T^{-1}A^T T$, $A^T = TAT^{-1}$):

$$A^T P + PA - PBB^T P + C^T C = 0 \Rightarrow$$

$$TAT^{-1}P + PA - PBCT^{-1}P + TBC = 0$$

Multiplying left by matrix T^{-1} we get:

$$AT^{-1}P + T^{-1}PA - T^{-1}PBCT^{-1}P + BC = 0$$

So making $P^* = T^{-1}P$

$$AP^* + P^*A - P^*BCP^* + BC = 0 \tag{9.15}$$

Proceeding analogously with the FARE equation (9.1):

$$A\Pi + \Pi A^T - \Pi C^T C\Pi + BB^T = 0 \Rightarrow$$

$$A\Pi + \Pi TA^T T^{-1} - \Pi TBC\Pi + BCT^{-1} = 0$$

Multiplying right by T and making $P^* = \Pi T$ we get:

$$AP^* + P^*A - P^*BCP^* + BC = 0 \tag{9.16}$$

Relation (9.16), as you can see, equals equation (9.15). Therefore, analogous to open-loop balancing, CARE and FARE are equal for symmetrical systems. The obtained equation is still quadratic but in contrast to the Riccati equation the solution is not symmetrical. Furthermore, there is no guarantee that the solution is positive definite. Note however that:

$$P^* = \Pi T, P^* = T^{-1}P \Rightarrow P^* \cdot P^* = \Pi P \tag{9.17}$$

where Π and P are solutions for the FARE and CARE equations. If, out of all the solutions, the positive definite matrices $\bar{\Pi}$ and $\bar{P}$ are selected, they are able to define the characteristic values (or rather the square roots of the eigenvalues of the matrix product $\bar{\Pi}\bar{P}$). So, for symmetrical systems and because of relation (9.17), the characteristic values can be calculated as the square roots of the eigenvalues of matrix $(P^*)^2$. Since matrix T can be calculated from the observability and controllability matrix, to determine the characteristic values, instead of resolving two Riccati equations we only need to resolve one.

So, equation (9.15) is equivalent to the cross-gramian equation and is called the cross-Riccati equation. Analogous to what happened with the W_{co} matrix, P^* also holds two pieces of information: one is that it can calculate the system characteristic values and furthermore, the eigenvalue signs allow it to determine the signature matrix of the system.

9.6 Exercises

1. Given matrix

$$A = \begin{bmatrix} 0 & 1 & 0 & 0 \\ 0 & 0 & 1 & 0 \\ 0 & 0 & 0 & 1 \\ -3 & -8 & -8 & -5 \end{bmatrix} \tag{9.18}$$

   determine $\|\Delta A\|_{max}$ so that $A + \Delta A$ is a matrix with eigenvalues with negative real part.

2. Given the system with transfer function

$$G(s) = \frac{1}{(1-s)(s+2)^2(s+0.5)^2}$$

   determine a balanced closed-loop realization and an opportune reduced order model. Then verify that the model is closed-loop stable.

3. Calculate the characteristic values for the system with transfer function $G(s) = \frac{2s}{s^2+1}$.

4. Calculate the closed-loop balanced realization for the system with transfer function $G(s) = \frac{s+1}{s^5+7s^2+6s+5}$. Design a reduced order regulator and observer making sure that the closed-loop system is asymptotically stable.

5. Given the system with transfer function $G(s) = \frac{s}{s^2+1} + \frac{s^2+1}{s(s^2+2)}$ calculate the characteristic values and synthesize the system with circuit components.

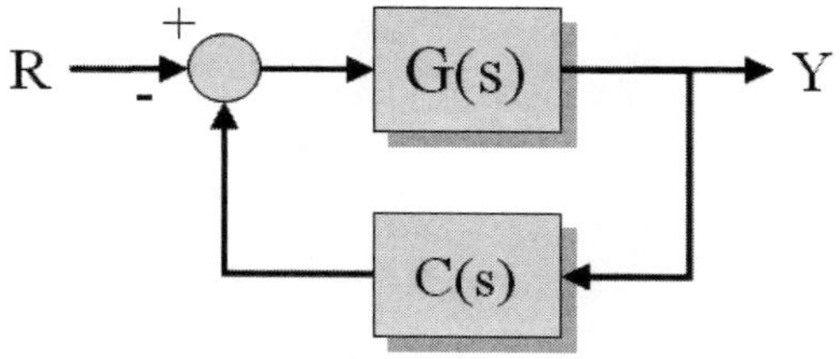

FIGURE 9.4
Block scheme for exercise 7.

6. Calculate, if possible, the system that has characteristic values $\rho_1 = 5$ e $\rho_2 = 1$ and eigenvalues $\lambda_1 = -5$ and $\lambda_2 = -1$.
7. Given the system with transfer function $G(s) = \frac{1}{s-1}$ calculate the linear state regulator, optimal observer and the compensator $C(s)$ (see Figure 9.4), using the CARE and FARE equations.
8. Verify that the characteristic values of an all-pass system are all equal to k.
9. Design a reduced order model for system $G(s) = \frac{s+1}{s^3+5s^2+7s-2}$.
10. Given system

$$G(s) = \frac{1}{(1-s)(s+2)^2(s+0.5)^2}$$

determine a closed-loop balanced realization and a reduced order model. Verify then that the reduced order model is stable in closed-loop.

10

Passive and bounded-real systems

CONTENTS

In this chapter the main properties of passive systems and bounded-real systems are discussed. To the reader the duality between them will appear.

10.1 Passive systems

There is perfect duality between dynamical systems and circuits. A dynamical system can be made of circuits which can mimic its behavior. Consider the circuit in Figure 10.1 with parameters $L > 0$, $C > 0$, $R_1 > 0$, $R_2 > 0$. This type of circuit is passive: it dissipates the energy supplied by the generator without ever being able to supply any. Making up the circuit is well known: in particular, by attributing an equivalent impedance to the inductor and capacitor, the circuit can be studied according in the Laplace domain (see Figure 10.2). Thus, working with algebraic equations rather than with

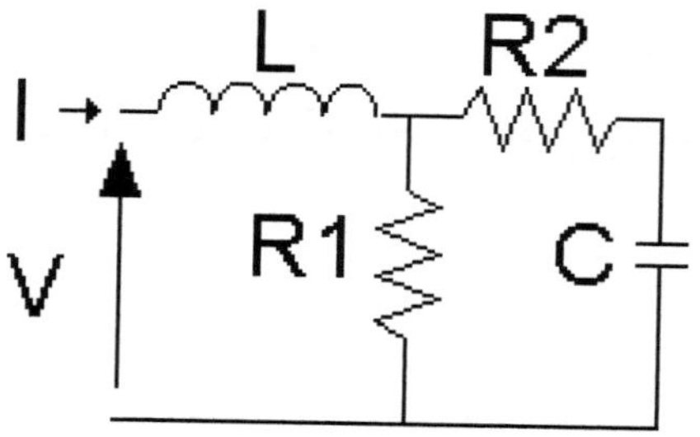

FIGURE 10.1
Example of a passive electrical circuit ($L > 0$, $C > 0$, $R_1 > 0$, $R_2 > 0$).

integral-differential equations, the relationships characterizing the circuit can be derived. For example, suppose the impedance of a circuit is defined as

$$Z(s) = \frac{V(s)}{I(s)} \tag{10.1}$$

Calculating this characteristic transfer function is easy. More generally, when a circuit has more than one input (e.g., the circuit in Figure 10.3 has two independent current generators), the circuit can be characterized by a matrix of impedances, each from each input to each output.

First, let's consider what conditions make a SISO system passive, then extending this to systems with multiple inputs and outputs.

10.1.1 Passivity in the frequency domain

Deciding whether a SISO system is passive is done by analyzing its impedance characteristics by means of the concept of positive-real function.

Definition 16 (Positive-real function) *A function is positive-real if it satisfies the following properties:*

1. function $Z(s)$ is analytical for $\Re e\ s > 0$ (i.e., there are no poles on the right-hand half of the complex plane);

2. any function poles lying on the imaginary axis are simple and have positive and real residues;;

3. $\Re e\ (Z(j\omega)) \geq 0\ \forall \omega$.

Once a positive-real function is defined, we discuss how to determine if the system is passive by analyzing its impedance characteristics. Let's look at the following theorem:

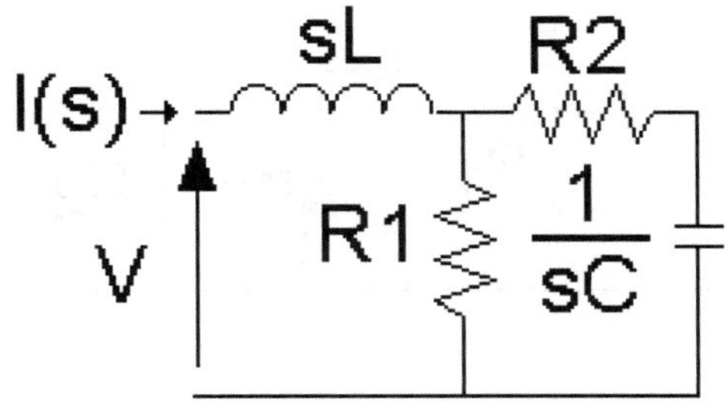

FIGURE 10.2
Example of a passive electrical circuit ($L > 0$, $C > 0$, $R_1 > 0$, $R_2 > 0$) studied in the Laplace domain.

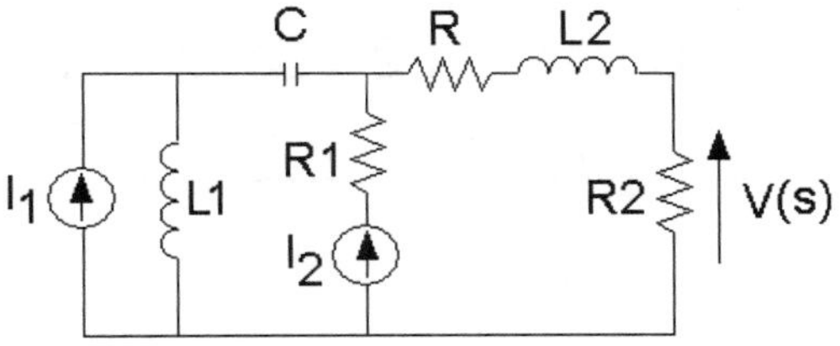

FIGURE 10.3
Example of a passive electrical circuit with two inputs.

Theorem 17 *A SISO system is passive if and only if its impedance is positive-real.*

The importance of the last condition of the positive-real definition should be highlighted; the fact that the real part of the impedance is positive is clearly linked to the fact that the system is dissipating energy. Next, we will speak equivalently about positive-real systems and passive ones. Furthermore, strictly positive-real systems will refer to those systems strictly satisfying property 3) in definition 16, i.e., $\Re e\ (Z(j\omega)) > 0\ \forall\omega$.

Let's look at some examples of classes of passive systems. Firstly, generic RLC systems (here and in the next, we assume $R > 0$, $L > 0$ and $C > 0$) have a positive-real impedance which makes them passive.

Another example of passive systems are positive-real functions with simple poles and positive residues whose impedance derives from circuits made up

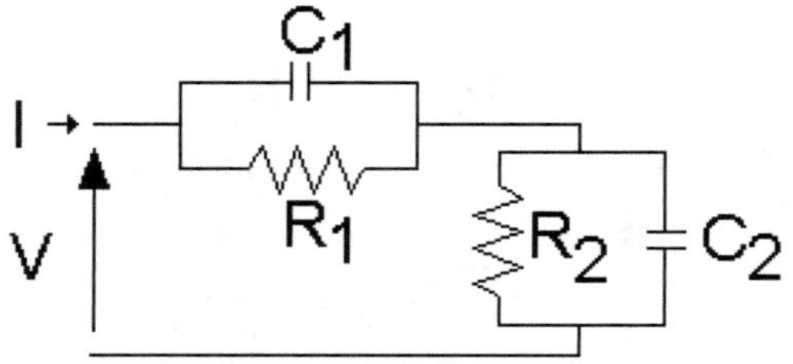

FIGURE 10.4
Example of a relaxation electrical circuit.

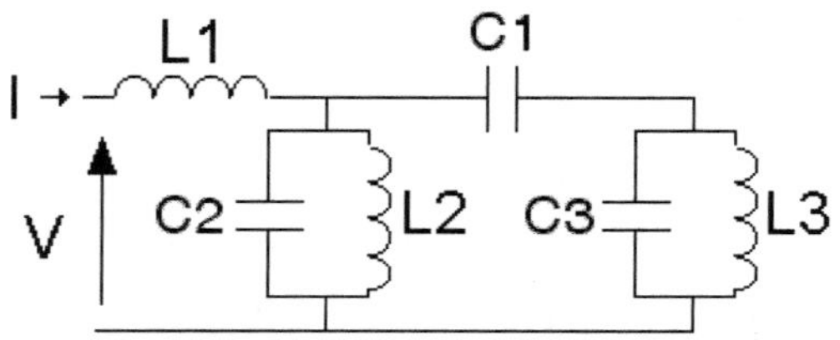

FIGURE 10.5
Example of a *loss-less* circuit.

only of resistors and capacitors, the so-called class of *relaxation systems*. An example is shown in Figure 10.4.

In *loss-less* systems, i.e., circuits made up of ideal capacitors and inductors (an example is shown in Figure 10.5) the impedance is always such that $\Re e\ (Z(j\omega)) = 0\ \forall \omega$ (precisely because there are no losses). This impedance is odd positive-real ($Z(s) = -Z(-s)$), its particularity being that its poles and zeros (all on the imaginary axis, otherwise it would degenerate or dissipate energy) alternate in frequency.

Example 10.1

Let's consider the systems whose pole-zero maps are shown in Figure 10.6:

- $Z(s) = \frac{s^2+1}{s(s^2+4)}$ (Figure 10.6(a)) is a positive-real loss-less system;
- $Z(s) = \frac{s}{s^2+1}$ (Figure 10.6(b)) is a positive-real loss-less system;
- $Z(s) = \frac{s^2+4}{s^2+1}$ (Figure 10.6(c)) is not a positive-real loss-less system;
- $Z(s) = \frac{s^2+4}{s(s^2+1)}$ (Figure 10.6(d)) is not a positive-real loss-less system.

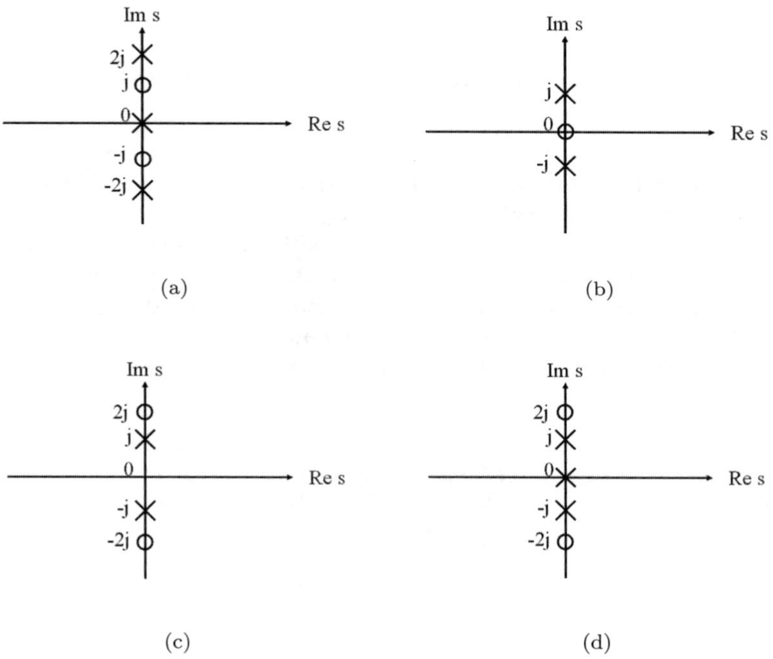

FIGURE 10.6
Examples of systems having (a) and (b) and not having (c) and (d) the property of losslessness.

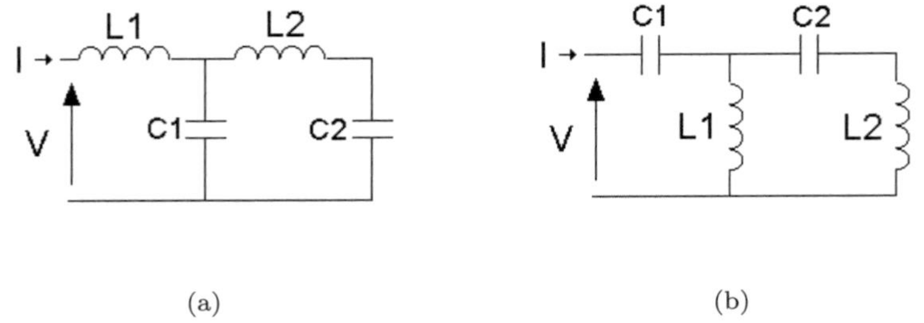

FIGURE 10.7
Examples of loss-less positive-real systems: (a) low-pass filter; (b) high-pass filter.

Note that, relaxation systems are strictly odd positive-real, whereas loss-less systems are positive-real but not strictly so.

The characteristic values of an odd positive-real function are all one, whereas the singular values cannot be calculated given that Lyapunov equations have no solution. This means that a lower order model cannot be proposed which is based on closed-loop balancing since all the state variables have equal weight in the feedback and so have to be necessarily taken into consideration.

Example 10.2

Let's calculate the characteristics of an odd positive-real function $G(s) = \frac{2}{s}$. Consider the realization: A = 0, B = 1, C = 2. The CARE equation is:

$$-P^2 + 4 = 0$$

and so $\bar{P} = 2$. The FARE equation is:

$$-4\Pi^2 + 1 = 0$$

and so $\bar{\Pi} = \frac{1}{2}$.
Therefore

$$\bar{\Pi}\bar{P} = 1 \Rightarrow \mu = 1$$

Example 10.3

Let's consider the two circuits in Figure 10.7. The one in Figure 10.7(a) is a low-pass filter, whereas the one in Figure 10.7(b) is a high-pass filter. Since both circuits are loss-less, their characteristic values are one.

Note that this result also holds for MIMO systems.

Also note that the term *positive systems* refers to a different class of systems. A linear system is said to be externally positive only if its output (the zero-state response) is non-negative for whatever non-negative input.

Theorem 18 *A linear system is externally positive if and only if its impulse response is non-negative.*

An as-yet open question is how to verify the non-negativity of $y(t) = \mathrm{C}e^{\mathrm{A}t}\mathrm{B}$, the impulse response of a continuous linear time-invariant system. This requires numerical verification except when there are special pole and residue properties as in the case of relaxation systems which are externally positive (having positive residues).

Analogous to the SISO systems, passive MIMO systems are those whose transfer matrix is positive-real.

Definition 17 (Positive-real transfer matrix) *A transfer matrix is positive-real if and only if:*

1. *each of its components is analytical for $\Re e\ s > 0$ (i.e., there are no poles on the right-hand half of the complex plane);*
2. *for each of the imaginary poles of $\mathrm{Z}(s)$, if any, its residue is a positive semi-definite Hermitian matrix;*
3. *$\mathrm{Z}(j\omega) + \mathrm{Z}^T(-j\omega)$ is positive semi-definite for each ω such that $j\omega$ is not a pole of $\mathrm{Z}(j\omega)$.*

Remember that $\mathrm{Z}^\dagger(j\omega)$ is the Hermitian matrix (i.e., the transposed conjugated matrix) of $\mathrm{Z}(j\omega)$, condition that $\mathrm{Z}^\dagger(j\omega) + \mathrm{Z}(j\omega)$ is positive definite $\forall\omega$ substitutes condition 4) of the positive-real definition. It implies that as ω varies, matrix $\mathrm{Z}^\dagger(j\omega) + \mathrm{Z}(j\omega)$ has only positive eigenvalues.

10.1.2 Passivity in the time domain

The definition discussed above is in the frequency domain. There are also analogous conditions to characterize passive systems in the time domain. Let's consider a realization $(\mathrm{A}, \mathrm{B}, \mathrm{C}, \mathrm{D})$ of the impedance matrix $\mathrm{Z}(s)$, assumed square (i.e., $m = p$). To characterize a passive system from a realization of it, the following theorem, known as the Positive-Real Lemma, is needed.

Theorem 19 (Positive-Real Lemma) *A linear time-invariant system $\mathrm{R}(\mathrm{A}, \mathrm{B}, \mathrm{C}, \mathrm{D})$ is positive-real if there exists a positive definite matrix P which satisfies:*

$$\begin{cases} \mathrm{PA} + \mathrm{A}^T\mathrm{P} = -\mathrm{LL}^T \\ \mathrm{PB} = \mathrm{C}^T - \mathrm{LW} \\ \mathrm{D} + \mathrm{D}^T = \mathrm{W}^T\mathrm{W} \end{cases} \tag{10.2}$$

with $\mathrm{L} \in \mathbb{R}^{n\times m}$ and $\mathrm{W} \in \mathbb{R}^{m\times m}$.

Note in particular that the factorization of equation (10.2) exists if matrix D is positive definite (this in fact allows to obtain a factorizable matrix $\mathrm{D}+\mathrm{D}^T$ as the product of two matrices.

Consider, for example, an improper system. It decomposes into a strictly proper part and a remainder. From a circuitry point of view, these two terms correspond to a dynamical circuit block (being a function of s) and a resistance. If the resistance is negative, the circuit is not passive and the function is not positive-real. This corresponds exactly to when D is not positive definite.

On the hypothesis that matrix D is positive definite, L can be calculated from the second equation of (10.2):

$$\mathrm{L} = (-\mathrm{PB} + \mathrm{C}^T)\mathrm{W}^{-1} \tag{10.3}$$

Note that matrix W is invertible since $\mathrm{D}+\mathrm{D}^T \neq 0$, given that D is positive definite.

For this reason, it is not enough that D is positive semi-definite, but that D is required to be definite positive.

By substituting into the first (10.2), equation we obtain:

$$\mathrm{PA} + \mathrm{A}^T\mathrm{P} = -(-\mathrm{PB} + \mathrm{C}^T)(\mathrm{D} + \mathrm{D}^T)^{-1}(-\mathrm{PB} + \mathrm{C}^T)^T \tag{10.4}$$

and by re-ordering

$$\begin{aligned} \mathrm{P}(\mathrm{A} - \mathrm{B}(\mathrm{D} + \mathrm{D}^T)^{-1}\mathrm{C}) + (\mathrm{A}^T - \mathrm{C}^T(\mathrm{D} + \mathrm{D}^T)^{-1}\mathrm{B}^T)\mathrm{P} + \\ + \mathrm{PB}(\mathrm{D} + \mathrm{D}^T)^{-1}\mathrm{B}^T\mathrm{P} + \mathrm{C}^T(\mathrm{D} + \mathrm{D}^T)^{-1}\mathrm{C} = 0 \end{aligned} \tag{10.5}$$

Equation (10.5) is a Riccati equation in which the quadratic term is positive so there is no guarantee of a positive definite solution. A Hamiltonian matrix can be associated with Ricatti equation (10.5):

$$\mathrm{H} = \begin{bmatrix} \mathrm{A} - \mathrm{B}(\mathrm{D} + \mathrm{D}^T)^{-1}\mathrm{C} & \mathrm{B}(\mathrm{D} + \mathrm{D}^T)^{-1}\mathrm{B}^T \\ -\mathrm{C}^T(\mathrm{D} + \mathrm{D}^T)^{-1}\mathrm{C} & -\mathrm{A}^T + \mathrm{C}^T(\mathrm{D} + \mathrm{D}^T)^{-1}\mathrm{B}^T \end{bmatrix} \tag{10.6}$$

According to theorem 19, a system is positive-real if Riccati equation (10.5) provides a definite positive solution.

The real positiveness of a system can be also studied by examining the Hamiltonian matrix (10.6). The condition can be derived by taking into account the following theorem relating the eigenvalues of the Hamiltonian matrix (10.6) with the singularity of the matrix $\mathrm{Z}(j\omega) + \mathrm{Z}^T(-j\omega)$ at some ω.

Theorem 20 *Assume that* A *has no imaginary eigenvalues,* $(\mathrm{D}+\mathrm{D}^T)$ *is non-singular and* $\omega_0 \in \mathbb{R}$*. Then* $\lambda = 0$ *is an eigenvalue of* $(\mathrm{Z}(j\omega_0) + \mathrm{Z}^T(-j\omega_0))$ *if and only if* $(\mathrm{H} - j\omega_0\mathrm{I})$ *is singular.*

For MIMO systems with $n = 1$ and for SISO systems it can be derived that the system is positive-real if the Hamiltonian matrix (10.6) has no eigenvalues on the imaginary axis.

10.1.3 Factorizing positive-real functions

Having obtained matrix P, matrix L can also be determined, and from these the matrix

$$\mathrm{W}(s) = \mathrm{W} + \mathrm{L}^T(s\mathrm{I} - \mathrm{A})^{-1}\mathrm{B}$$

which factorizes the original positive-real transfer matrix as follows:

$$\mathrm{Z}(s) + \mathrm{Z}^T(-s) = \mathrm{W}^T(-s)\mathrm{W}(s)$$

Therefore, a positive-real transfer matrix can be factorized as the product of two matrices which depend on s.

Example 10.4

Consider, for example, $Z(s) = \frac{2}{s}$ (positive-real function).
Since $Z^T(-s) = -\frac{2}{s}$, then $W(s) = 0$ is capable of factorizing the original positive-real function.

Example 10.5

Now consider $Z(s) = -\frac{2}{s}$ which is not a positive-real function. Consider the realization: $A = 0$, $B = 1$ and $C = -2$ (negative residue). Consider too, the second equation of the Positive-Real Lemma (10.2): $PB = C^T - LW$. Given that $C < 0$, no P value can satisfy this equation. So, neither L nor $W(s)$ can be calculated.

10.1.4 Passification

Given a not passive system, one may ask whether it is possible to make it passive by a feedforward action. This problem referred to as the problem of *passification* is dealt with in this section. It is worth noticing that this can be done for MIMO stable systems and only requires a single feedforward parameter as proven in the following theorem. Essentially, the theorem states that a MIMO stable system $\mathrm{G}(s) = \left[\begin{array}{c|c} \mathrm{A} & \mathrm{B} \\ \hline \mathrm{C} & \mathrm{D} \end{array}\right]$, in general not passive, can be made passive through the addition of $\alpha\mathrm{I}$ to its feedforward matrix D.

Theorem 21 *Let* $\mathrm{G}(s) = \left[\begin{array}{c|c} \mathrm{A} & \mathrm{B} \\ \hline \mathrm{C} & \mathrm{D} \end{array}\right]$ *be a minimal realization of an asymptotically stable system, then there exists* $\alpha \geq 0$ *such that the system* $\mathrm{G}_\alpha(s) = \left[\begin{array}{c|c} \mathrm{A} & \mathrm{B} \\ \hline \mathrm{C} & \mathrm{D} + \alpha\mathrm{I} \end{array}\right]$ *is passive.*

Proof 6 *If* $\mathrm{G}(s)$ *is passive, trivially* $\alpha = 0$ *and* $\mathrm{G}_\alpha(s)$ *is also passive, so that, in the following, we consider the case that* $\mathrm{G}(s)$ *is not passive.*

Let us define the following quantity:

$$\bar{\lambda} := \inf_{\omega} \lambda_{min}(\mathrm{G}(j\omega) + \mathrm{G}^T(-j\omega)) \tag{10.7}$$

Since $\mathrm{G}(s)$ *is not passive,* $\bar{\lambda} < 0$*. Moreover, since* $G(s)$ *is a proper asymptotically stable transfer matrix, it has no poles on the imaginary axis and* $\bar{\lambda} > -\infty$*. Therefore,* $\bar{\lambda}$ *is a real finite quantity.*

Let us define as $\bar{\omega}$ *the point at which* $\lambda_{min}(\mathrm{G}(j\bar{\omega}) + \mathrm{G}^T(-j\bar{\omega})) = \bar{\lambda}$ *(in the limit case, it can be also* $\bar{\omega} = +\infty$*).*

Let us now consider the matrix $(\mathrm{G}(j\bar{\omega}) + \mathrm{G}^T(-j\bar{\omega}))$*. This is an Hermitian (and therefore diagonalizable) matrix:*

$$\mathrm{G}(j\bar{\omega}) + \mathrm{G}^T(-j\bar{\omega}) = \mathrm{T}\Lambda\mathrm{T}^T \tag{10.8}$$

where T *is an unit matrix containing the orthogonal eigenvector of* $\mathrm{G}(j\bar{\omega}) + \mathrm{G}^T(-j\bar{\omega})$ *and* $\Lambda = diag\{\lambda_1, \lambda_2, \ldots, \lambda_n\}$*. The smallest eigenvalue of* $\mathrm{G}(j\bar{\omega}) + \mathrm{G}^T(-j\bar{\omega})$ *is* $\bar{\lambda}$*.*

If the quantity $\alpha\mathrm{I}$ *is added to the* D *matrix of the original system* $\mathrm{G}(s)$*, one obtains:*

$$\begin{aligned} &\mathrm{G}_\alpha(j\bar{\omega}) + \mathrm{G}_\alpha^T(-j\bar{\omega}) = \mathrm{G}(j\bar{\omega}) + \mathrm{G}^T(-j\bar{\omega}) + 2\alpha\mathrm{I} = \\ &= \mathrm{T}\Lambda\mathrm{T}^T + 2\alpha\mathrm{I} = \mathrm{T}(\Lambda + 2\alpha\mathrm{I})\mathrm{T}^T \end{aligned} \tag{10.9}$$

Therefore, if α *is chosen as* $\alpha \geq -\bar{\lambda}/2$*, all the eigenvalues of* $\mathrm{G}_\alpha(j\bar{\omega}) + \mathrm{G}_\alpha^T(-j\bar{\omega})$ *are non-negative, from which it can be concluded that the original system may be made passive by adding* $\alpha\mathrm{I}$ *to the original* D *matrix. Moreover, if* $\alpha > -\bar{\lambda}/2$*, the system* G_α *is strictly passive.*

The result can be also extended to Lyapunov stable systems.

The minimum value of α, say $\bar{\alpha}$, that makes the system passive can be found checking the eigenvalues of the Hamiltonian matrix:

$$\mathrm{H}_\alpha = \begin{bmatrix} \mathrm{A} - \mathrm{B}(\mathrm{D} + \mathrm{D}^T + 2\alpha\mathrm{I})^{-1}\mathrm{C} & \mathrm{B}(\mathrm{D} + \mathrm{D}^T + 2\alpha\mathrm{I})^{-1}\mathrm{B}^T \\ -\mathrm{C}^T(\mathrm{D} + \mathrm{D}^T + 2\alpha\mathrm{I})^{-1}\mathrm{C} & -\mathrm{A}^T + \mathrm{C}^T(\mathrm{D} + \mathrm{D}^T + 2\alpha\mathrm{I})^{-1}\mathrm{B}^T \end{bmatrix} \tag{10.10}$$

associated to the system having in place of matrix D matrix $\tilde{\mathrm{D}} = \mathrm{D} + \alpha\mathrm{I}$. This is rigorously defined for all α such that $\alpha \neq -\lambda_i((\mathrm{D} + \mathrm{D}^T)/2)$, i.e., $-\alpha$ should not be an eigenvalue of the symmetrical part of D. $\bar{\alpha}$ is that value that guarantees that the minimum eigenvalue $\lambda_{min}(\omega)$ of $\mathrm{G}_\alpha(j\omega) + \mathrm{G}_\alpha^T(-j\omega)$ is positive.

Let us consider a value of α, say $\alpha*$, such that $\lambda_{min}(\omega)$ crosses the real axis (i.e., the corresponding H_α has imaginary eigenvalues). For $\alpha \geq \alpha*$, $\mathrm{G}_\alpha(s)$ is passive if and only if H_α has no eigenvalues on the imaginary axis. Therefore, the following steps can be adopted to find the minimum value of α such that $\mathrm{G}_\alpha(s)$ is passive. Pick a random value of ω, say ω_0, and let $\alpha_l = -\lambda_{min}(\mathrm{G}(j\omega_0) + \mathrm{G}^T(-j\omega_0))/2$. Let then α_h be a real number large enough that H_{α_h} has no eigenvalues on the imaginary axis (i.e., such that $G_{\alpha_h}(s)$ is passive). In summary, the following bisection algorithm has been used to find the minimum value of α such that $G_\alpha(s)$ is passive:

- fix $\alpha = (\alpha_l + \alpha_h)/2$;

- calculate H_α;
- if H_α has no imaginary eigenvalues, then fix $\alpha_h = \alpha$, otherwise $\alpha_l = \alpha$;
- stop the procedure when $\alpha_h - \alpha_l < \epsilon$, where $\epsilon > 0$ is the required precision for the calculation of the minimum value of α.

Example 10.6

Let us consider a SISO system defined by the following state-space matrices:

$$A = \begin{bmatrix} 0 & 1 \\ -5 & -5 \end{bmatrix}; B = \begin{bmatrix} 0 \\ 1 \end{bmatrix}; C = \begin{bmatrix} 7 & -9 \end{bmatrix}; D = 1 \tag{10.11}$$

The transfer function of system (10.11) is $G(s) = \frac{s^2-4s+12}{s^2+5s+5}$. Since, for SISO systems, $G(j\omega)+G^T(-j\omega) = 2\text{Re}\,[G(j\omega)]$, the real part of $G(j\omega)$ has to be studied. In particular, the values at which $\text{Re}\,[G(j\omega)] = 0$ can be calculated from the characteristic polynomial of the Hamiltonian matrix H. This is given by:

$$\psi(s) = \det(sI - H) = s^4 + 37s^2 + 60 \tag{10.12}$$

Solving for $\psi(j\omega) = 0$ yields two positive solutions: $\omega_1 = 1.3037$ and $\omega_2 = 5.9414$. For these values, we thus have $\text{Re}\,[G(j\omega)] = 0$. In fact, $\text{Re}\,[G(j\omega)]$ is given by:

$$\text{Re}\,[G(j\omega)] = \frac{\omega^4 - 37\omega^2 + 60}{(5-\omega^2)^2 + 25\omega^2} \tag{10.13}$$

The plot of this function is shown in Figure 10.8.

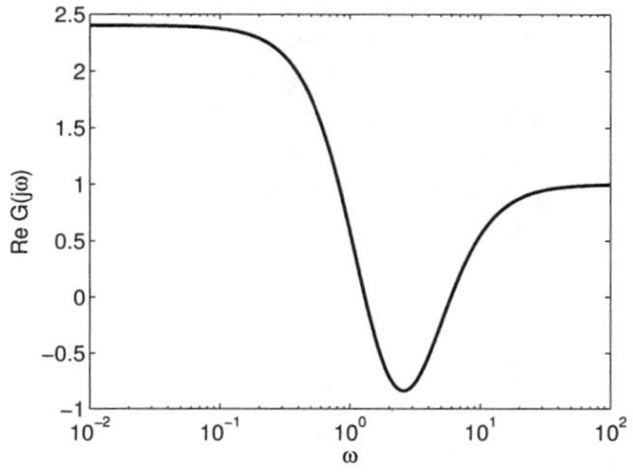

FIGURE 10.8
Plot of Re $[G(j\omega)]$ vs. ω for example 10.6.

Solving for $\frac{d\text{Re}\,[G(j\omega)]}{d\omega} = 0$ (i.e., $\frac{\omega(104\omega^4-140\omega^2-3650)}{((5-\omega^2)^2+25\omega^2)^2} = 0$) yields $\bar{\omega} = 2.5759$. Since $\text{Re}\,[G(j\bar{\omega})] = -0.8394$, then $\bar{\lambda} = -1.6788$. Therefore, α has to be chosen so that $\alpha \geq \frac{\bar{\lambda}}{2} = 0.8394$. For instance, the system

$$\begin{array}{ll} A = \begin{bmatrix} 0 & 1 \\ -5 & -5 \end{bmatrix}; & B = \begin{bmatrix} 0 \\ 1 \end{bmatrix}; \\ C = \begin{bmatrix} 7 & -9 \end{bmatrix}; & D = 1 + \alpha; \\ \alpha = 0.8394 & \end{array} \tag{10.14}$$

is passive, since $\text{Re}\,[G_\alpha(j\omega)] \geq 0\ \forall\omega$ as shown in Figure 10.9.

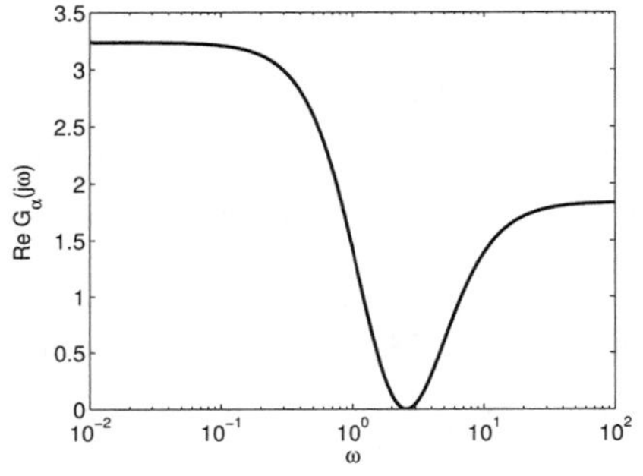

FIGURE 10.9
Plot of Re $[G_\alpha(j\omega)]$ for system (10.14) vs. ω.

MATLAB® exercise 10.1

Let us consider the MIMO system defined by the following state-space matrices:

$$A = \begin{bmatrix} 0 & 1 & 0 & 0 \\ 0 & 0 & 1 & 0 \\ 0 & 0 & 0 & 1 \\ -1 & -2 & -5 & -2 \end{bmatrix}; \quad B = \begin{bmatrix} 1 & 2 \\ -1 & 1 \\ 0 & 1 \\ -1 & 1 \end{bmatrix}; \quad C = \begin{bmatrix} 1 & -1 & 1 & -1 \\ 1 & 1 & 1 & 1 \end{bmatrix}; \quad D = \begin{bmatrix} 1.5 & 2 \\ 1 & 4 \end{bmatrix}. \tag{10.15}$$

The eigenvalues of $G(j\omega) + G^T(-j\omega)$ are shown in Figure 10.10. The corresponding eigenvalues of H are: $\lambda_{1,2}(H) = \pm 3.4219$, $\lambda_{3,4}(H) = \pm j0.5945$, $\lambda_{5,6}(H) = \pm j1.0602$, $\lambda_{7,8}(H) = \pm j1.6629$. These values indicate that the eigenvalues of $G(j\omega) + G^T(-j\omega)$ cross the real axis three times, i.e., in correspondence of $\omega_1 = 0.5945$, $\omega_2 = 1.0602$ and $\omega_3 = 1.6629$. In this case, it is the minimum eigenvalue of $G(j\omega) + G^T(-j\omega)$ that crosses the real axis at ω_1, ω_2 and ω_3.

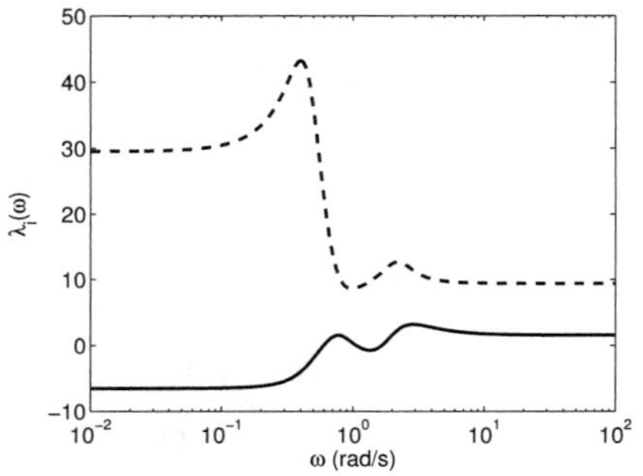

FIGURE 10.10
Eigenvalues of $G(j\omega) + G^T(-j\omega)$ for system (10.15).

By applying the bisection algorithm with $\epsilon = 0.0001$, it can be found that the minimum value of α such that $G(s)$ is passive is $\alpha = 3.2674$.

In correspondence of this value, the eigenvalues of H are: $\lambda_{1,2,3,4}(H) = \pm 0.7699 \pm j1.5088$, $\lambda_{5,6}(H) = \pm 0.4873$ and $\lambda_{7,8}(H) = \pm 0.0040$. The eigenvalues of $G_\alpha(j\omega) + G_\alpha^T(-j\omega)$ for $\alpha = 3.2674$ are shown in Figure 10.11.

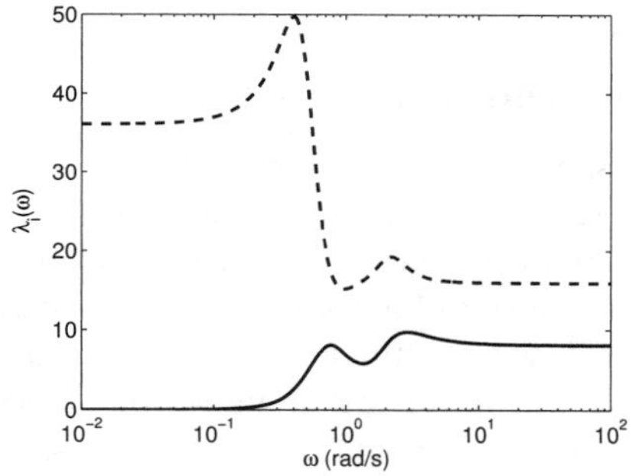

FIGURE 10.11
Eigenvalues of $G_\alpha(j\omega) + G_\alpha^T(-j\omega)$ for system (10.15) with $\tilde{D} = D + \alpha I$ and $\alpha = 3.2674$.

The MATLAB® code used is here reported:

```
A=[0 1 0 0; 0  0 1 0; 0 0 0 1; -1 -2 -5 -2];
B=[1 2; -1 1; 0 1; -1 1];
C=[1 -1 1 -1; 1 1 1 1];
D=[1.5 2; 1 4];
system=ss(A,B,C,D)
MH=[A-B*inv(D+D')*C B*inv(D+D')*B';
  -C'*inv(D+D')*C -A'+C'*inv(D+D')*B'];
eig(MH)
w=logspace(-2,2,1000);
autovaloreminvsomega=w;
E=[];
Z=freqresp(system,w);
ZZ=Z;
for i=1:length(w)
    ZZ=Z(:,:,i)+ctranspose(Z(:,:,i));
    autovaloreminvsomega(i)=min(eig(ZZ));
    E=[E'; eig(ZZ)']';
end
figure,semilogx(w,E)
alpha=3.2674;
D=alpha*eye(2)+system.D;
systemalpha=ss(A,B,C,D);
MH=[A-B*inv(D+D')*C B*inv(D+D')*B';
  -C'*inv(D+D')*C -A'+C'*inv(D+D')*B'];
eig(MH)
w=logspace(-2,2,1000);
autovaloreminvsomega=w;
E=[];
Z=freqresp(systemalpha,w);
ZZ=Z;
for i=1:length(w)
    ZZ=Z(:,:,i)+ctranspose(Z(:,:,i));
```

```
    autovaloreminvsomega(i)=min(eig(ZZ));
    E=[E'; eig(ZZ)']';
end
figure,semilogx(w,E)
```

10.1.5 Passive reduced order models

Now let's consider the issue of finding a reduced order model of a positive-real system to obtain. Suppose the system has no poles on the imaginary axis, and to calculate the open-loop balanced realization from which to construct a lower-order model, once the singular values have been calculated. To evaluate the efficacy of the model, it is not sufficient to consider the error norm between the original system and the lower-order model but since this latter should accurately describe the original system characteristics, it must be verified it is still a positive-real system.

There is a method which guarantees that the lower-order model is still positive-real. As we saw in Chapter 6, it is important to build a lower-order model into a state-space realization which has the desired characteristics. In this case, consider the dual equation of the Riccati equation (10.4):

$$\Pi \mathrm{A}^T + \mathrm{A}\Pi = -(-\Pi \mathrm{C}^T + \mathrm{B})(\mathrm{D} + \mathrm{D}^T)^{-1}(-\Pi \mathrm{C}^T + \mathrm{B})^T \tag{10.16}$$

and find a realization in which the solutions to (10.4) and (10.16) are equal and diagonal. Naming these solutions $\bar{\Pi}$ and $\bar{\mathrm{P}}$ note they are positive definite matrices given the original system is positive-real. Furthermore, since they are diagonal, when direct truncation is applied and the matrix subblock neglected, the subblock obtained is positive definite so the original characteristics of the system (i.e., passivity) are preserved. When dealing with symmetrical systems, the lower-order model can be obtained via one equation.

10.1.6 Energy considerations connected to the Positive-Real Lemma

The Positive-Real Lemma came about as a result of considerations on circuit energy. Let's consider the impedance in Figure 10.12. Given that the current and tension flows are as in Figure 10.12, the energy dissipated by the impedance is:

$$E = \int_0^T v(t)i(t)dt \tag{10.17}$$

The circuit is passive if the dissipated energy is not negative. For a linear time-invariant system:

$$\begin{cases} \dot{\mathbf{x}} = \mathrm{A}\mathbf{x} + \mathrm{B}\mathbf{u} \\ \mathbf{y} = \mathrm{C}\mathbf{x} + \mathrm{D}\mathbf{u} \end{cases} \tag{10.18}$$

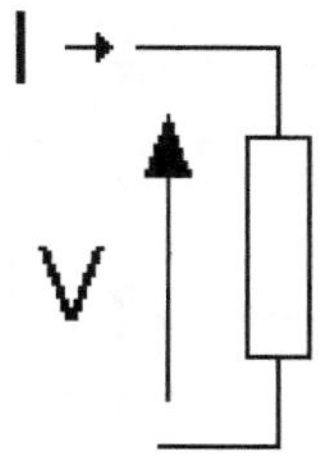

FIGURE 10.12
Impedance.

analogous to equation (10.12) the energy can be defined as:

$$E = \int_0^T \mathbf{u}^T(t)\mathbf{y}(t)dt \tag{10.19}$$

System (10.18) is passive if energy (10.19) is non-negative. This energy condition means that the inequalities must apply:

$$\mathrm{P} > 0; \quad \begin{bmatrix} \mathrm{A}^T\mathrm{P} + \mathrm{PA} & \mathrm{PB} - \mathrm{C}^T \\ \mathrm{B}^T\mathrm{P} - \mathrm{C} & -\mathrm{D}^T - \mathrm{D} \end{bmatrix} \leq 0 \tag{10.20}$$

Equation (10.20) is a collection of matrix inequalities; if they have a solution then the system is passive. In Chapter 12 we will see that many control problems can be re-formulated as linear matrix inequalities.

10.1.7 Closed-loop stability and positive-real systems

There is a significant connection between the stability of feedback systems and positive-real systems. Let's consider the feedback system in Figure 10.13, it can be shown that if $G_1(s)$ and $G_2(s)$ are positive-real and at least one of the two systems is strictly positive-real, then the closed-loop system is asymptotically stable.

Theorem 22 *The feedback system shown in Figure 10.13 is asymptotically stable if $G_1(s)$ e $G_2(s)$ are positive-real, at least one of the two systems being strictly positive-real.*

Example 10.7

Consider the system in Figure 10.13 with $G_1(s) = \frac{1}{s+4}$ and $G_2(s) = \frac{s}{s^2+1}$. Since $G_2(s)$ is loss-less (therefore positive-real) and $G_1(s)$ is a relaxation system (therefore strictly positive-real) the closed-loop system is asymptotically stable. It is easy to verify that

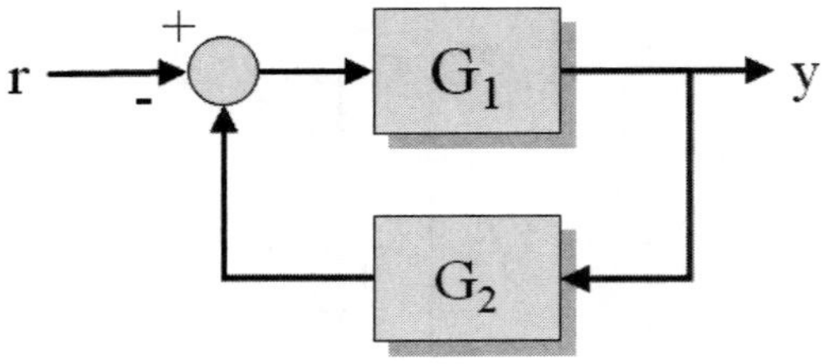

FIGURE 10.13
Closed-loop system with positive-real $G_1(s)$ and $G_2(s)$.

the closed-loop transfer function equals $W(s) = \frac{G_1(s)}{1+G_1(s)G_2(s)} = \frac{s^2+1}{s^3+4s^2+2s+4}$ and that the closed-loop system is asymptotically stable.

Example 10.8

The condition expressed by the theorem is only a sufficient condition, the opposite not being verified. Let's look at two counter-examples of two systems which do not satisfy the theorem conditions, one case showing a closed-loop unstable system, the other asymptotically stable.
First, let's consider $G_1(s) = \frac{1}{s-1}$ and $G_2(s) = \frac{s}{s^2+1}$. The closed-loop transfer function is $W(s) = \frac{s^2+1}{s^3-s^2+2s-1}$ and the closed-loop system is unstable.
Now let's consider $G_1(s) = \frac{1}{s-0.1}$ and $G_2(s) = \frac{1}{s^2+2s+1}$, here the closed-loop transfer function is $W(s) = \frac{s-0.1}{s^3+1.9s^2+0.8s+0.9}$. By calculating the system poles and applying the Routh criteria the closed-loop system can be verified as being asymptotically stable.

10.1.8 Optimal gain for loss-less systems

There is a significant tie-in between optimal gain and matrix C coefficients for the loss-less class of systems. Consider the optimal control problem for a loss-less system $S(\mathrm{A}, \mathrm{B}, \mathrm{C})$, given the following optimization index:

$$J = \int_0^\infty (x^T \mathrm{C}^T \mathrm{C} x + u^T u) dt \tag{10.21}$$

Optimal gain turns out to be equal to matrix C, i.e., $\mathrm{K} = \mathrm{C}$.

A system $S(\mathrm{A}, \mathrm{B}, \mathrm{C})$ is loss-less if and only if the Positive-Real Lemma holds, in other words if there exists a symmetrical and positive definite P_1 matrix such that:

$$\mathrm{P}_1 \mathrm{A} + \mathrm{A}^T \mathrm{P}_1 = 0 \tag{10.22}$$

$$\mathrm{P}_1 \mathrm{B} = \mathrm{C}^T \tag{10.23}$$

Now, let's consider the CARE equation for loss-less systems:

$$A^T P + PA - PBB^T P + C^T C = 0 \tag{10.24}$$

The solution of this equation allows optimal gain to be calculated, but by using relations (10.22) and (10.23), it can be proved that P_1 satisfies CARE and therefore $K = B^T P_1 = C$.

Moreover, it follows another result concerning the optimal eigenvalues for SISO loss-less systems with index (10.21): the optimal eigenvalues are the roots of

$$N(s) + D(s) = 0 \tag{10.25}$$

where $N(s)$ and $D(s)$ are the numerator and denominator polynomials of the transfer function of system $S(A, B, C)$.

MATLAB® exercise 10.2

Consider the continuous-time LTI system with state-space matrices:

$$A = \begin{bmatrix} 0 & 1 & 0 & 0 \\ 0 & 0 & -1 & 0 \\ 0 & 0 & 0 & 8 \\ 0 & 0 & -8 & 0 \end{bmatrix}; \quad B = \begin{bmatrix} 1 \\ 0 \\ 0 \\ 1 \end{bmatrix}; \quad C = \begin{bmatrix} 2 & 0 & 0 & 4 \end{bmatrix}$$

The transfer function of this system is:

$$G(s) = \frac{6s^3 + 132s}{s^4 + 65s^2 + 64}$$

from which it is immediate to verify that the system is loss-less ($G(s) = -G^T(-s)$). Now, with the help of MATLAB, we solve the Riccati equation for optimal control and verify that $K = C$.

First, let us define the state-space matrices:

```
>> A=[0 1 0 0; -1 0 0 0; 0 0 0 8; 0 0 -8 0]
>> B=[1 0 0 1]'
>> C=[2 0 0 4]
```

and calculate its transfer function:

```
>> system=ss(A,B,C,0)
>> tf(system)
```

Let us now solve the Riccati equation $A^T P + PA - PBB^T P + C^T C$ associated with index (10.21). To do this, use the command:

```
>> [K,P,E]=lqr(A,B,C'*C,1)
```

One gets:

$$K = \begin{bmatrix} 2 & 0 & 0 & 4 \end{bmatrix}$$

i.e., $K = C$.

The corresponding optimal eigenvalues are given by: $\lambda_{1,2} = -1.8456 \pm i7.2052$, $\lambda_3 = -1.5738$ and $\lambda_4 = -0.7351$. They can be also calculated by equation (8.34) through the command:

```
>> roots([1 6 65 132 64])
```

Also note that equation (10.25) leads to the same result of equation (8.34).

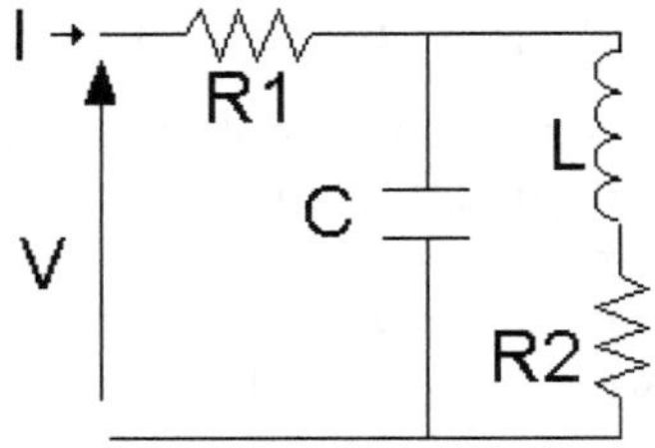

FIGURE 10.14
For $R1 = 1k\Omega$, $C = 2\mu F$, $L = 1mH$ and $R2 = 5k\Omega$ the passive circuit implements the transfer function $G(s) = \frac{2\cdot 10^{-6}s^2+(10+10^{-3})s+6\cdot 10^3}{2\cdot 10^{-9}s^2+10^{-2}s+1}$.

10.2 Circuit implementation of positive-real systems

In this section an example of the implementation of positive-real systems is illustrated.

Example 10.9

Given the system with transfer function $G(s) = \frac{2\cdot 10^{-6}s^2+(10+10^{-3})s+6\cdot 10^3}{2\cdot 10^{-9}s^2+10^{-2}s+1}$ propose, if possible, a passive electrical circuit implementation.

Solution

First verify if function $G(s)$ is positive-real, i.e., if it is possible to made a passive electric circuit which is analogous to the given system. To do this, we verify if conditions 1-3 of definition 16 are true. Points 1-2 can be checked immediately. For point 3, the real part of $G(j\omega)$ needs to be calculated:

$$\Re e(G(j\omega)) = \frac{6\cdot 10^3 + (10^{-1} - 4\cdot 10^{-6})\omega^2 + 4\cdot 10^{-15}\omega^4}{(1-2\cdot 10^{-9}\omega^2)^2 + 10^{-4}\omega^2}$$

Note that $\Re e\ (G(j\omega)) \geq 0\ \forall\omega$, so the system is positive-real.
Proceed by rewriting the transfer function as the sum of two terms, one strictly proper and one improper. To do this, we divide the polynomial numerator by the denominator, obtaining:

$$G(s) = 1000 + \frac{10^{-3}s + 5\cdot 10^3}{2\cdot 10^{-9}s^2 + 10^{-2}s + 1}$$

The first term can be interpreted as a resistor (R1) of value $1k\Omega$, while the second term is a generic impedance in series with R1. Rewrite the second term as follows:

$$G(s) = 1000 + \frac{1}{\frac{2\cdot 10^{-9}s^2+10^{-2}s+1}{10^{-3}s+5\cdot 10^3}}$$

and iterate the procedure with the term $\frac{2\cdot 10^{-9}s^2+10^{-2}s+1}{10^{-3}s+5\cdot 10^3}$. Note that this term should be interpreted as the admittance associated with the impedance in series with R1. Divide the polynomial numerator by the denominator to obtain:

$$G(s) = 1000 + \frac{1}{2 \cdot 10^{-6} s + \frac{1}{10^{-3} s + 5 \cdot 10^3}}$$

This form of $G(s)$ permits an immediate interpretation in terms of series and parallel impedances or elementary admittances. In fact the term $10^{-3}s + 5 \cdot 10^3$ can be interpreted as the series impedance of an inductor with a resistance ($L = 1mH$ and $R2 = 5k\Omega$). The series of these two components should be put in parallel with an admittance equal to $2 \cdot 10^{-6}s$ (and so with a capacitor $C = 2\mu F$). Finally, we have to consider the resistor R1 in series to this circuit block. The equivalent circuit is shown in Figure 10.14.

10.3 Bounded-real systems

Before defining bounded-real systems in the more general case, consider a SISO system. A SISO system $G(s)$ is bounded-real if $G(s)$ is stable and if the magnitude Bode diagram is below the value 0dB for any ω, that is if maximum gain is less than or equal to one.

From this definition it follows that $G(s) = \frac{1}{s}$ and $G(s) = \frac{4}{s+1}$ are not bounded-real systems, while $G(s) = \frac{1}{s+2}$ is.

Generally, the following defines bounded-real systems.

Definition 18 *A system* $\mathrm{S}(s)$ *is bounded-real if these conditions apply:*

1. *all the transfer matrix elements, i.e.,* $S_{ij}(s)$*, are analytical in* $\Re e\ s > 0$ *(i.e., the polynomials in the denominator of* $S_{ij}(s)$ *are Hurwitz polynomials);*

2. *matrix* $\mathrm{I} - \mathrm{S}^T(-s)\mathrm{S}(s)$ *is a non-negative Hermitian matrix for* $\Re e\ s > 0$ *or matrix* $\mathrm{I} - \mathrm{S}^T(-j\omega)\mathrm{S}(j\omega)$ *is non-negative for all the values of* ω*.*

Condition 2) can also be formulated according to the standard definition of the H_∞ norm of a system.

Remember that H_∞ norm of a system $\mathrm{S}(s)$ is defined as

$$||\mathrm{S}(s)||_\infty = \max_\omega[\sigma_{\max}(\mathrm{S}(j\omega))]$$

Condition 2) can be thus written as

$$||\mathrm{S}(s)||_\infty \leq 1$$

Moreover the system is called strictly bounded-real if $||\mathrm{S}(s)||_\infty < 1$.

Example 10.10

Consider system $G(s) = \frac{1}{s+2}$; condition 1) is clearly satisfied. Condition 2) implies that

$$1 - \frac{1}{2-j\omega}\frac{1}{2+j\omega} < 1$$

i.e., that

$$\frac{1}{4+\omega^2} < 1$$

which is the same condition obtained by applying the definition of bounded-real systems to SISO systems.

As in the case of real positive systems, conditions can be given directly in the time domain through the bounded-real lemma.

Theorem 23 (Bounded-real lemma) *A system* $\mathrm{S}(s) = \left[\begin{array}{c|c} \mathrm{A} & \mathrm{B} \\ \hline \mathrm{C} & \mathrm{D} \end{array}\right]$ *is bounded-real if* $\exists$ P *symmetric and positive definite that satisfies*

$$\mathrm{PA} + \mathrm{A}^T\mathrm{P} = -\mathrm{C}^T\mathrm{C} - \mathrm{LL}^T \tag{10.26}$$

$$-\mathrm{PB} = \mathrm{C}^T\mathrm{D} + \mathrm{LW} \tag{10.27}$$

$$\mathrm{I} - \mathrm{D}^T\mathrm{D} = \mathrm{W}^T\mathrm{W} \tag{10.28}$$

with $\mathrm{L} \in \mathbb{R}^{n\times m}$ *and* $\mathrm{W} \in \mathbb{R}^{m\times m}$.

Provided that W is invertible, we can write

$$\mathrm{L} = (-\mathrm{PB} - \mathrm{C}^T\mathrm{D})\mathrm{W}^{-1}$$

and substituting in the condition (10.26), we obtain the quadratic equation associated with the bounded-real condition of a system:

$$\begin{array}{l} \mathrm{P}(\mathrm{A} + \mathrm{B}(\mathrm{I} - \mathrm{D}^T\mathrm{D})^{-1}\mathrm{D}^T\mathrm{C}) + (\mathrm{A}^T + \mathrm{C}^T\mathrm{D}(\mathrm{I} - \mathrm{D}^T\mathrm{D})^{-1}\mathrm{B}^T)\mathrm{P} + \\ +\mathrm{PB}(\mathrm{I} - \mathrm{D}^T\mathrm{D})^{-1}\mathrm{B}^T\mathrm{P} + \mathrm{C}^T\mathrm{D}(\mathrm{I} - \mathrm{D}^T\mathrm{D})^{-1}\mathrm{D}^T\mathrm{C} + \mathrm{C}^T\mathrm{C} = 0 \end{array} \tag{10.29}$$

If the system is strictly proper, (D = 0), then equation (10.29) becomes

$$\mathrm{PA} + \mathrm{A}^T\mathrm{P} + \mathrm{PBB}^T\mathrm{P} + \mathrm{C}^T\mathrm{C} = 0 \tag{10.30}$$

Equation (10.29) can be associated with the following Hamiltonian matrix:

$$\mathrm{H} = \left[\begin{array}{cc} \mathrm{A} + \mathrm{B}(\mathrm{I} - \mathrm{D}^T\mathrm{D})^{-1}\mathrm{D}^T\mathrm{C} & \mathrm{B}(\mathrm{I} - \mathrm{D}^T\mathrm{D})^{-1}\mathrm{B}^T \\ -\mathrm{C}^T\mathrm{D}(\mathrm{I} - \mathrm{D}^T\mathrm{D})^{-1}\mathrm{D}^T\mathrm{C} - \mathrm{C}^T\mathrm{C} & -\mathrm{A}^T - \mathrm{C}^T\mathrm{D}(\mathrm{I} - \mathrm{D}^T\mathrm{D})^{-1}\mathrm{B}^T \end{array}\right] \tag{10.31}$$

If there exists a positive definite solution of the Riccati equation (10.29), then the system is bounded-real, otherwise not. Analogous to positive-real systems, the real boundness of a system can be checked from the eigenvalues of H. For strictly proper MIMO systems, for MIMO systems with $n = 1$, and for SISO systems, the condition is that H has no eigenvalues on the imaginary axis.

In strictly proper systems (D = 0) the Hamiltonian matrix (10.31) can be written thus:

$$\mathrm{H} = \begin{bmatrix} \mathrm{A} & \mathrm{B}\mathrm{B}^T \\ -\mathrm{C}^T\mathrm{C} & -\mathrm{A}^T \end{bmatrix} \tag{10.32}$$

Example 10.11

Consider system $G(s) = \frac{k}{s+1}$, taking into account the Bode diagram, the system is bounded-real if $k \leq 1$, and strictly bounded-real if $k < 1$.
Applying the bounded-real lemma, we obtain the same result. In fact a realization in state form of the system is given by A $= -1$, B $= 1$, C $= k$, for which the Hamiltonian is:

$$\mathrm{H} = \begin{bmatrix} -1 & 1 \\ -k^2 & 1 \end{bmatrix}$$

The characteristic polynomial is given by:

$$\det(\lambda \mathrm{I} - \mathrm{H}) = \lambda^2 - 1 + k^2 = 0$$

The condition that the eigenvalues are not on the imaginary axis is $|k| \leq 1$. For these values the system is bounded-real.
Note that if we consider the CARE associated with $G(s)$

$$\mathrm{H} = \begin{bmatrix} -1 & -1 \\ -k^2 & 1 \end{bmatrix}$$

we obtain

$$\det(\lambda \mathrm{I} - \mathrm{H}) = \lambda^2 - 1 - k^2 = 0$$

In this case, the eigenvalues of H are never purely imaginary (independent of the value assumed by k), so the issue of the optimal regulator is always solvable (since the system is always controllable).

If the system $\mathrm{S}(s)$ is bounded-real, i.e., if we are able to determine the positive definite solution of the Riccati equation (10.29), then $\mathrm{W}(s) = \mathrm{W} + \mathrm{L}^T(s\mathrm{I} - \mathrm{A})^{-1}\mathrm{B}$ can be defined which leads to the following factorization:

$$\mathrm{I} - \mathrm{S}^T(-s)\mathrm{S}(s) = \mathrm{W}^T(-s)\mathrm{W}(s)$$

The bounded-real lemma derives from a more general definition which can be given to dynamic systems (not necessarily linear). A system is bounded-real if $\forall u$, $\forall T$ is

$$\int_0^T y^T(t)y(t)dt \leq \int_0^T u^T(t)u(t)dt$$

The conditions expressed by the bounded-real lemma are limit conditions of this more general definition.

Finally, the problem can be reformulated in LMI terms. A system is

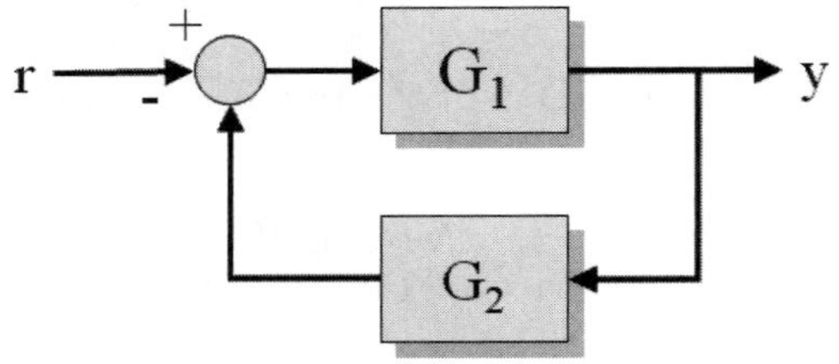

FIGURE 10.15
System constituted by the feedback of $G_1(s)$ and $G_2(s)$. The system is asymptotically stable if $G_1(s) \cdot G_2(s)$ is a strictly bounded-real system.

bounded-real if a positive definite solution among the solutions of the inequalities

$$\begin{array}{l} \mathrm{P} > 0 \\ \left[\begin{array}{cc} \mathrm{A}^T\mathrm{P} + \mathrm{PA} + \mathrm{C}^T\mathrm{C} & \mathrm{PB} + \mathrm{C}^T\mathrm{D} \\ \mathrm{B}^T\mathrm{P} + \mathrm{D}^T\mathrm{C} & \mathrm{D}^T\mathrm{D} - \mathrm{I} \end{array} \right] \leq 0 \end{array} \tag{10.33}$$

can be found (recall that the set of solutions of linear matrix inequalities is convex).

10.3.1 Properties of bounded-real systems

There is an important result linking the properties of bounded-real systems with closed-loop asymptotic stability expressed by the following theorem:

Theorem 24 (Small gain theorem) *The system shown in Figure 10.15 is asymptotically stable if $G_1(s) \cdot G_2(s)$ is a strictly bounded-real system.*

The proof of this theorem is based on the fact that, injecting a signal into the summing node, it will recur attenuated in input. In fact, since the system is bounded-real, each frequency component of the signal will be attenuated. Note also that in the SISO case the theorem implies that the Nyquist plot of the equation $L(s) = G_1(s)G_2(s)$ does not surround point -1.

The theorem expresses a very important result. Since the norm H_∞ is a consistent norm, we can obtain information about the stability of the feedback system from the value of the norm of the single systems. For example, if $\|G_1(s)\|_\infty < \gamma$ we can deduce that the feedback system will be asymptotically stable for any system $G_2(s)$ such that $\|G_2(s)\|_\infty < \frac{1}{\gamma}$. As we will see below, this result is fundamentally important for characterizing the robustness of a control system.

10.3.2 Bounded-real reduced order models

The Riccati equation (10.29) is also important for obtaining a reduced-order model with particular properties. If we approximate a bounded-real system using the open-loop balancing, in fact, we have no guarantee that the approximated system is bounded-real. We ought to have an approximated model that faithfully represents this characteristic of the initial system.

To approximate a bounded-real system through a reduced-order model which is also bounded-real, the Riccati equation (10.29) and its dual can be used for balancing and then proceed similarly to closed-loop balancing. In addition, if the system is symmetrical, the two equations can be replaced by a single Riccati equation.

10.4 Relationship between passive and bounded-real systems

There is a relationship between passive systems and bounded-real systems. Before clarifying this relationship, we will briefly show why we need to find a link between passive systems and bounded-real systems. The theory of passive filter design often requires a specification in the frequency domain which the designed filter must meet. This mask is defined by two curves in the Bode diagram which represent the upper and lower limits within which the transfer function of the designed filter must lie. Given transfer function $G(j\omega)$, it has then to be implemented through components R, L and C. If the transfer function is real positive, as we have seen, this implementation is always possible.

Assigned a mask, the problem of designing a filter is to find a transfer function which adheres to the specifications defined by the mask and at the same time is real positive. This problem is difficult to solve, whereas it is easier to find a bounded-real function that respects the specifics defined by the mask. The importance of a link between passive and bounded-real systems now becomes evident, because it permits us, once the bounded-real function is found, to obtain a passive system which satisfies the assigned specifics. This relation is expressed through a matrix called a "scattering matrix".

If $\mathrm{S}(s)$ is a transfer matrix of a bounded-real system, a scattering matrix can be defined as:

$$\mathrm{G}(s) = [\mathrm{I} + \mathrm{S}(s)][\mathrm{I} - \mathrm{S}(s)]^{-1}$$

which has the positiveness property.

Moreover, if $\mathrm{S}(s) = \left[\begin{array}{c|c} \tilde{\mathrm{A}} & \tilde{\mathrm{B}} \\ \hline \tilde{\mathrm{C}} & \tilde{\mathrm{D}} \end{array}\right]$, then a realization of $\mathrm{G}(s)$ is given by:

$$
\mathrm{G}(s)=\left[\begin{array}{c|c}(\tilde{\mathrm{A}}+\mathrm{I})(\tilde{\mathrm{A}}-\mathrm{I})^{-1} & \sqrt{2}(\tilde{\mathrm{A}}-\mathrm{I})^{-1}\tilde{\mathrm{B}} \\ \hline -\sqrt{2}(\tilde{\mathrm{A}}^T-\mathrm{I})^{-1}\tilde{\mathrm{C}} & \tilde{\mathrm{D}}-\tilde{\mathrm{C}}^T(\tilde{\mathrm{A}}-\mathrm{I})^{-1}\tilde{\mathrm{B}}\end{array}\right]
$$

10.5 Exercises

1. Given the continuous-time system with transfer function $G(s) = \frac{\alpha}{s^3+s^2+4s+4}$ calculate for which values of α is the system bounded-real.
2. Given the continuous-time system with transfer function $G(s) = \frac{4}{s^2+\alpha s+4}$ calculate for which values of α is the system bounded-real.
3. Given the continuous-time system with state-space matrices:

$$
\mathrm{A}=\begin{bmatrix} 0 & 1 & 0 \\ 0 & 0 & 1 \\ -5 & -3 & -2 \end{bmatrix}; \quad \mathrm{B}=\begin{bmatrix} 0 \\ 0 \\ \alpha \end{bmatrix} \quad \mathrm{C}=\begin{bmatrix} 1 & 0 & 0 \end{bmatrix}
$$

calculate for which values of α is the system bounded-real.

4. Given the continuous-time system with state-space matrices:

$$
\mathrm{A}=\begin{bmatrix} -1 & 0 & 0 \\ 0 & -2 & 0 \\ 0 & 0 & -0.5 \end{bmatrix}; \quad \mathrm{B}=\mathrm{C}^T=\begin{bmatrix} \alpha \\ \alpha \\ \alpha \end{bmatrix}
$$

calculate for which values of α is the system strictly bounded-real.

5. For the feedback system shown in Figure 10.16 with $G(s) = \frac{1}{s^2}$ prove, if possible, that there is a stable first order compensator which makes the system closed-loop passive. Note: do not make any cancellations.
6. Given the system with transfer function $G(s) = \frac{s+1}{(s+2)(s+3)}$ calculate the Cauchy index of the system with two different analytical methods. Then calculate the energy associated with the impulse response. Finally, using the bounded-real lemma, verify analytically if $G(s)$ is bounded-real.
7. Determine for which values of α the system with transfer function $G(s) = \frac{\alpha}{2s^3+s^2+4s+5}$ is bounded-real using two different methods.
8. Given the system with transfer function $G(s) = \frac{s(s^2+2)}{(s^2+1)(s^2+4)}$, propose, if possible, a passive electric circuit realization.

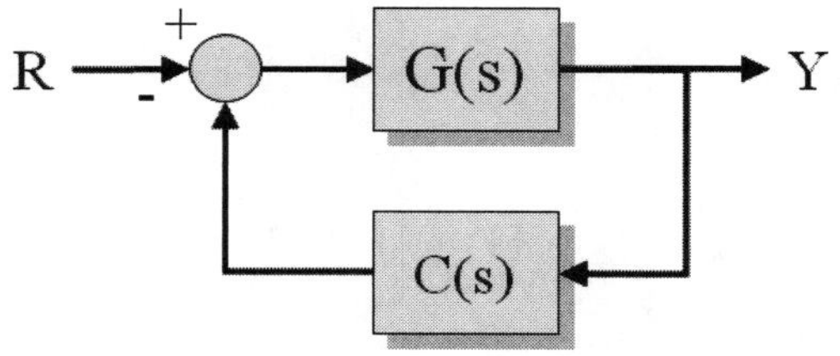

FIGURE 10.16
Block scheme for exercise 5.

9. Given the system $G(s) = \frac{1}{(s+1)^2}$ determine if there is an R such that the optimal closed-loop system (with functional $J = \int_0^\infty (y^T y + u^T Ru)dt$) is passive.
10. Write down an example of a third-order loss-less system and verify that its characteristic values are equal to one.

11

H_∞ linear control

CONTENTS

11.1 Introduction

As we know, the transfer functions which characterize the classic control scheme (shown in Figure 11.1) are:

$$\frac{Y(s)}{D(s)} = \frac{1}{1 + C(s)P(s)H(s)} \tag{11.1}$$

$$\frac{Y(s)}{R(s)} = \frac{C(s)P(s)}{1 + C(s)P(s)H(s)} \tag{11.2}$$

$$\frac{Y(s)}{N(s)} = -\frac{C(s)P(s)H(s)}{1 + C(s)P(s)H(s)} \tag{11.3}$$

The analysis of these transfer functions reveals that some of the typical requirements of a control system involve specific conflicts. For example, the requirement of input tracking requires a large bandwidth, but this implies a deterioration in performance in terms of rejection of the measurement noise $n(t)$ within the bandwidth so that noise is not very attenuated. Conversely, the requirements of noise $d(t)$ rejection, of parametric insensitivity and input tracking are not in conflict with each other.

Given these considerations, in relation to the highlighted control system, it makes sense to consider the problem of finding the parameters of $C(s)$ which first assure the stability of the closed-loop system while minimizing the sensitivity of the transfer function $S(s) = \frac{Y(s)}{D(s)}$, thus assuring good disturbance attenuation for all the bandwidth frequencies. Since the minimizing index must account for disturbance varying with frequency, it can be represented by the H_∞ norm of the sensitivity function.

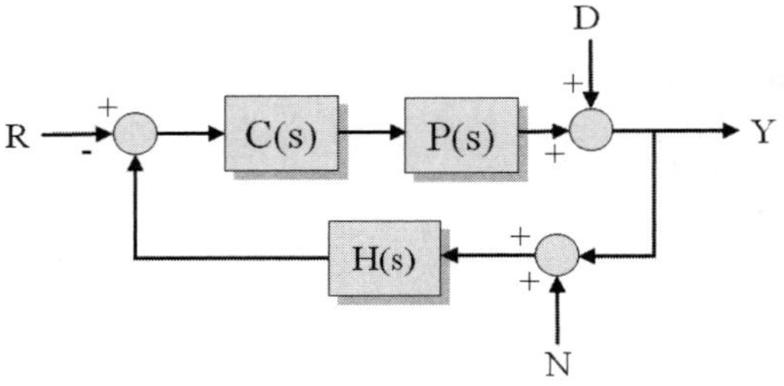

FIGURE 11.1
Scheme of classical control.

H_∞ control allows us to solve this kind of problem and more: a general control scheme which allows us to treat MIMO systems and more general problems is considered.

Before illustrating what is meant by H_∞ control, let's discuss how to calculate the H_∞ norm of a system. The definition of H_∞ norm, in previous chapters, does not allow the calculation, so we need to use an iterative algorithm. In fact, there is no closed formula to calculate this norm.

Consider system $G(s)$, the fact that the H_∞ norm is less than γ ($\|G(s)\|_\infty < \gamma$) is equivalent to the fact that the system $G(s)/\gamma$ is bounded real. Basically, the idea is to calculate the H_∞ norm based on this observation. Since it is possible to determine whether a system is bounded real or not through the Bounded Real Lemma, calculating the H_∞ norm can be done by performing a Bounded Real Lemma test.

Given system $G(s) = (\mathrm{A}, \mathrm{B}, \mathrm{C})$, the H_∞ norm can be found, verifying for which values of γ the system $G(s)/\gamma = (\mathrm{A}, \frac{\mathrm{B}}{\gamma}, \mathrm{C})$ is bounded real. So we have to consider the Hamiltonian matrix

$$\mathrm{H} = \begin{bmatrix} \mathrm{A} & \frac{\mathrm{BB}^T}{\gamma^2} \\ -\mathrm{C}^T\mathrm{C} & -\mathrm{A}^T \end{bmatrix} \tag{11.4}$$

fix a value of γ and verify if matrix (11.4) has no eigenvalues on the imaginary axis. If it does, system $G(s)/\gamma$ is not bounded real and the norm of $G(s)$ is not less than γ. At this point we have to choose a larger value of γ and run the test again. The H_∞ norm value is given by the smallest value of γ which assures that the system $G(s)/\gamma$ is still bounded real, that is the value for which the Hamiltonian matrix (11.4) has no eigenvalues on the imaginary axis.

Example 11.1

Consider the system with transfer function $G(s) = \frac{10}{s+5}$. From the Bode diagram we see that the H_∞ norm of this system is $\|G(s)\|_\infty = 2$. If we consider a realization of the system with $A = -5$; $B = 10$; $C = 1$ we can calculate the H_∞ norm using the procedure based on the Hamiltonian (11.4). In this case we obtain:

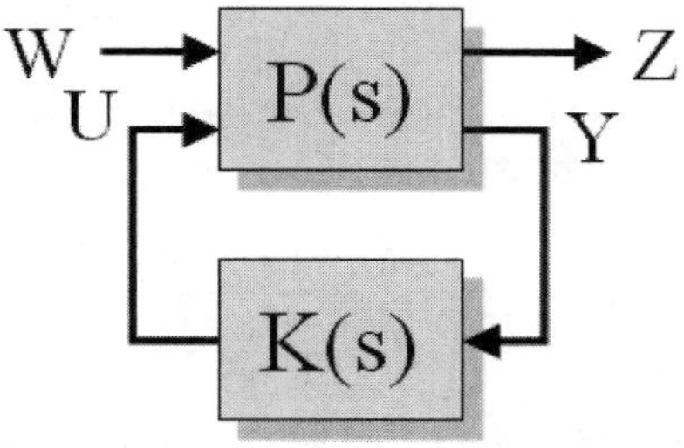

FIGURE 11.2
Scheme of robust control.

$$H = \begin{bmatrix} -5 & \frac{100}{\gamma^2} \\ -1 & 5 \end{bmatrix} \tag{11.5}$$

The characteristic polynomial of the Hamiltonian is $p(\lambda) = \lambda^2 - 25 + \frac{100}{\gamma^2}$, from which it can be derived that the limit condition for which the Hamiltonian has no eigenvalues on the imaginary axis is $\gamma = 2$.

11.2 Solution of the H_∞ linear control problem

The scheme used for H_∞ control is shown in Figure 11.2. P(s) shows the system to be controlled, K(s) is the compensator to be determined. The system outputs are represented by output variables available for the control (variables **y**) and by the interest variables (variables **z**). The objective of the control is to minimize the effect of the exogenous inputs (variables **w**) on the interest variables, acting through the manipulable inputs (or control signals), represented by the variables **u**.

The effect of the exogenous inputs (which include disturbances) on the variables of interest is represented by the transfer matrix from **w** to **z**, denoted by $\mathrm{T}_{zw}(s)$. We refer to the H_∞ norm as a measure of the size of the transfer matrix from **w** to **z**.

Once $\gamma > 0$ is assigned, the control problem H_∞ consists in determining the compensator K(s) which stabilizes the closed-loop system of Figure 11.2 and guarantees $\|\mathrm{T}_{zw}\|_\infty \leq \gamma$.

The state equations of the system P(s) in the H_∞ control scheme are given by:

$$\begin{cases} \dot{\mathbf{x}} = \mathrm{A}\mathbf{x} + \mathrm{B}_1\mathbf{w} + \mathrm{B}_2\mathbf{u} \\ \mathbf{z} = \mathrm{C}_1\mathbf{x} + \mathrm{D}_{11}\mathbf{w} + \mathrm{D}_{12}\mathbf{u} \\ \mathbf{y} = \mathrm{C}_2\mathbf{x} + \mathrm{D}_{21}\mathbf{w} + \mathrm{D}_{22}\mathbf{u} \end{cases} \tag{11.6}$$

where the dimensions of the vectors $\mathbf{x}$, $\mathbf{w}$, $\mathbf{u}$, $\mathbf{z}$ and $\mathbf{y}$ are respectively, n, m_1, m_2, p_1 and p_2.

With the compact notation given by the realization matrix the system to be controlled can be written as:

$$\mathrm{P}(s) = \left[\begin{array}{c|cc} \mathrm{A} & \mathrm{B}_1 & \mathrm{B}_2 \\ \hline \mathrm{C}_1 & \mathrm{D}_{11} & \mathrm{D}_{12} \\ \mathrm{C}_2 & \mathrm{D}_{21} & \mathrm{D}_{22} \end{array}\right]$$

The problem of the H_∞ control can be solved in the time domain or frequency domain. Originally, it was solved in 1988 in the frequency domain. Two years later, the problem was solved in the time domain, which permits correlating the results to those for optimal control and the Kalman filter.

Let's analyze in detail the time domain technique to solve the problem. This allows us to obtain a compensator of order n, defined by the two gain matrices, K_c for the regulator and K_e for the observer. Let's consider $\mathrm{D}_{11} = \mathrm{D}_{22} = 0$ and let's define as X_∞ and Y_∞ the two solutions of the Riccati equations expressed in compact form:

$$\mathrm{X}_\infty = \mathtt{RIC}\begin{bmatrix} \mathrm{A} - \mathrm{B}_2\tilde{\mathrm{D}}_{12}\mathrm{D}_{12}^T\mathrm{C}_1 & \gamma^{-2}\mathrm{B}_1\mathrm{B}_1^T - \mathrm{B}_2\tilde{\mathrm{D}}_{12}\mathrm{B}_2^T \\ -\tilde{\mathrm{C}}_1^T\tilde{\mathrm{C}}_1 & -(\mathrm{A} - \mathrm{B}_2\tilde{\mathrm{D}}_{12}\mathrm{D}_{12}^T\mathrm{C}_1)^T \end{bmatrix} \tag{11.7}$$

and

$$\mathrm{Y}_\infty = \mathtt{RIC}\begin{bmatrix} (\mathrm{A} - \mathrm{B}_1\mathrm{D}_{21}^T\tilde{\mathrm{D}}_{21}\mathrm{C}_2)^T & \gamma^{-2}\mathrm{C}_1^T\mathrm{C}_1 - \mathrm{C}_2^T\tilde{\mathrm{D}}_{21}\mathrm{C}_2 \\ -\tilde{\mathrm{B}}_1\tilde{\mathrm{B}}_1^T & -(\mathrm{A} - \mathrm{B}_1\mathrm{D}_{21}^T\tilde{\mathrm{D}}_{21}\mathrm{C}_2) \end{bmatrix} \tag{11.8}$$

with $\tilde{\mathrm{C}}_1 = (\mathrm{I} - \mathrm{D}_{12}\tilde{\mathrm{D}}_{12}\mathrm{D}_{12}^T)\mathrm{C}_1$ and $\tilde{\mathrm{B}}_1 = \mathrm{B}_1(\mathrm{I} - \mathrm{D}_{21}^T\tilde{\mathrm{D}}_{21}\mathrm{D}_{21})$.

Equation (11.7) is a compact way of indicating the solution of the Riccati equation:

$$\mathrm{A}^*\mathrm{X}_\infty + \mathrm{X}_\infty\mathrm{A}^{*T} + \mathrm{X}_\infty\mathrm{R}^*\mathrm{X}_\infty + \mathrm{Q}^* = 0$$

with $\mathrm{A}^* = \mathrm{A} - \mathrm{B}_2\tilde{\mathrm{D}}_{12}\mathrm{D}_{12}^T\mathrm{C}_1$, $\mathrm{R}^* = \gamma^{-2}\mathrm{B}_1\mathrm{B}_1^T - \mathrm{B}_2\tilde{\mathrm{D}}_{12}\mathrm{B}_2^T$ and $\mathrm{Q}^* = \tilde{\mathrm{C}}_1^T\tilde{\mathrm{C}}_1$. Similarly the equation (11.8) defines the solution of an analogue Riccati equation.

The solutions of these two Riccati equations depend on γ, which represents the objective reached by the H_∞ control. Starting from matrices X_∞ and Y_∞ the gains characterizing the compensator can be derived:

$$\begin{cases} \mathbf{K}_c = \tilde{\mathrm{D}}_{12}(\mathrm{B}_2^T\mathrm{X}_\infty + \mathrm{D}_{12}^T\mathrm{C}_1) \\ \mathbf{K}_e = (\mathrm{Y}_\infty\mathrm{C}_2^T + \mathrm{B}_1\mathrm{D}_{21}^T)\tilde{\mathrm{D}}_{21} \end{cases} \tag{11.9}$$

where

$$\begin{aligned} \tilde{\mathrm{D}}_{12} &= (\mathrm{D}_{12}^T\mathrm{D}_{12})^{-1} \\ \tilde{\mathrm{D}}_{21} &= (\mathrm{D}_{21}\mathrm{D}_{21}^T)^{-1} \end{aligned}$$

Note that the Hamiltonian associated with the Riccati equation (11.7), if γ is very large, coincides with the Hamiltonian for optimal control (so a solution can be found). The problem of the existence of a solution arises when γ is small, that is when the performance requirements are strong, in this case matrix $R^* = \gamma^{-2}B_1B_1^T - B_2\tilde{D}_{12}^T B_2^T$ may be non-negative definite.

The existence of a stabilizing compensator which guarantees $\|T_{zw}\|_\infty < \gamma$ is assured if there are two positive definite solutions to the Riccati equations (11.7) and (11.8) and if the maximum eigenvalue of the product matrix $X_\infty \cdot Y_\infty$ is less than γ^2, that is if $\lambda_{max}(X_\infty \cdot Y_\infty) < \gamma^2$.

The procedure for solving the H_∞ control problem can be summarized in the following steps:

1. determine a state-space representation for the process $P(s)$;
2. verify the existence conditions (invertibility of $D_{12}^T D_{12}$ and $D_{21}D_{21}^T$);
3. fix a positive value of γ large enough (to solve the two Riccati equations);
4. solve the two Riccati equations, obtaining the two positive definite solutions;
5. verify if the condition $\lambda_{max}(X_\infty \cdot Y_\infty) < \gamma^2$ is met;
6. if the steps 4) and 5) are verified, it is possible to repeat the procedure, lowering γ to point 3).

The obtained compensator consists of a control law expressed as $\mathbf{u} = -K_c\hat{\mathbf{x}}$ and an observer which dynamics is given by the following equation:

$$\dot{\hat{x}} = A\hat{\mathbf{x}} + B_2\mathbf{u} + B_1\hat{\mathbf{w}} + Z_\infty K_e(\mathbf{y} - \hat{\mathbf{y}}) \tag{11.10}$$

with $\hat{\mathbf{w}} = \gamma^{-2}B_1^T X_\infty$ and $\hat{\mathbf{y}} = C_2\hat{\mathbf{x}} + \gamma^{-2}D_{21}B_1^T X_\infty\hat{\mathbf{x}}$.

In compact form the state-space equations of the compensator are:

$$K(s) = \left[\begin{array}{c|c} A - B_2K_c - Z_\infty K_e C_2 + \gamma^{-2}(B_1B_1^T - Z_\infty K_e D_{21}B_1^T)X_\infty & Z_\infty K_e \\ \hline -K_c & 0 \end{array}\right]$$

with $Z_\infty = (I - \gamma^{-2}X_\infty \cdot Y_\infty)^{-1}$.

Example 11.2

As an example of H_∞ control consider the system in Figure 11.3. Suppose u is the torque applied by the motor to the axis and $J = 1$ is the moment of inertia of the axis. θ is the angle between the axis and x-axis. Assuming the engine is frictionless, the system model is represented by a double integrator. If x_1 is the angular velocity variable and x_2 the state variable representing the angular position, the model in state-space form is expressed by the following equations:

$$\begin{cases} \dot{x}_1 = d + u \\ \dot{x}_2 = x_1 \end{cases} \tag{11.11}$$

where we highlighted the presence of a disturbance d on the engine torque. The output

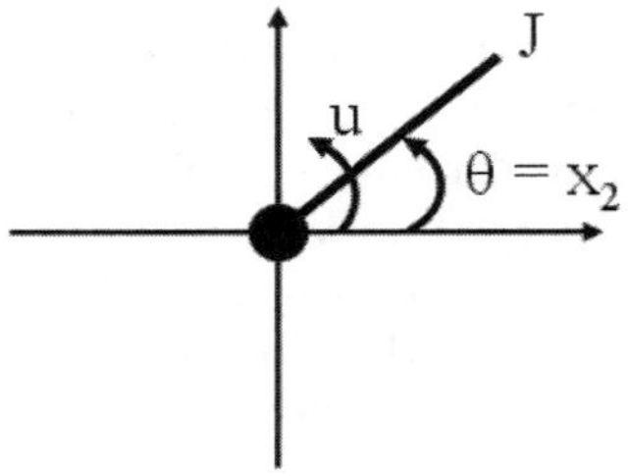

FIGURE 11.3
Example of system to be controlled with the H_∞ control.

to be regulated is defined as $z = \begin{bmatrix} x_2 \\ u \end{bmatrix}$, so as to take into account the effect of the disturbance on variable x_2, but also on input u. The output is given by variable x_2 affected by noise n ($y = x_2 + n$). Finally, the exogenous input vector is given by $w = \begin{bmatrix} d \\ n \end{bmatrix}$.
Summarizing, the complete model in state-space form is given by:

$$\begin{cases} \dot{x}_1 = w_1 + u \\ \dot{x}_2 = x_1 \\ z_1 = x_2 \\ z_2 = u \\ y = x_2 + w_2 \end{cases} \tag{11.12}$$

Recalling the formulation of state-space equations (11.6) for the H_∞ control scheme:

$$\begin{cases} \dot{\mathbf{x}} = \mathrm{A}\mathbf{x} + \mathrm{B}_1\mathbf{w} + \mathrm{B}_2\mathbf{u} \\ \mathbf{z} = \mathrm{C}_1\mathbf{x} + \mathrm{D}_{11}\mathbf{w} + \mathrm{D}_{12}\mathbf{u} \\ \mathbf{y} = \mathrm{C}_2\mathbf{x} + \mathrm{D}_{21}\mathbf{w} + \mathrm{D}_{22}\mathbf{u} \end{cases}$$

the realization matrix of the system to be controlled can be easily derived:

$$\mathrm{P}(s) = \left[\begin{array}{c|cc} \mathrm{A} & \mathrm{B}_1 & \mathrm{B}_2 \\ \hline \mathrm{C}_1 & \mathrm{D}_{11} & \mathrm{D}_{12} \\ \mathrm{C}_2 & \mathrm{D}_{21} & \mathrm{D}_{22} \end{array}\right] = \left[\begin{array}{cc|ccc} 0 & 0 & 1 & 0 & 1 \\ 1 & 0 & 0 & 0 & 0 \\ \hline 0 & 1 & 0 & 0 & 0 \\ 0 & 0 & 0 & 0 & 1 \\ 0 & 1 & 0 & 1 & 0 \end{array}\right]$$

Applying the H_∞ control procedure (shown in the next chapter), the transfer function of the compensator

$$K(s) = \frac{-977.4(s+0.40)}{(s+2.33)(s+373.4)}$$

and the optimum value of γ $\gamma_{ott} = 2.62$ are obtained. Finally, the solution matrices of the Riccati equations are $\mathrm{X}_\infty = \begin{bmatrix} 1.59 & 1.08 \\ 1.08 & 1.47 \end{bmatrix}$ and $\mathrm{Y}_\infty = \begin{bmatrix} 1.47 & 1.08 \\ 1.08 & 1.59 \end{bmatrix}$.

Example 11.3

In this exercise a further example of H_∞ control is given. Let us consider a Rapid Thermal Processing (RTP) system in semiconductor wafer manufacturing, in which the ability of rapidly changing the temperature is important for the fabrication of devices with very small crystal length.

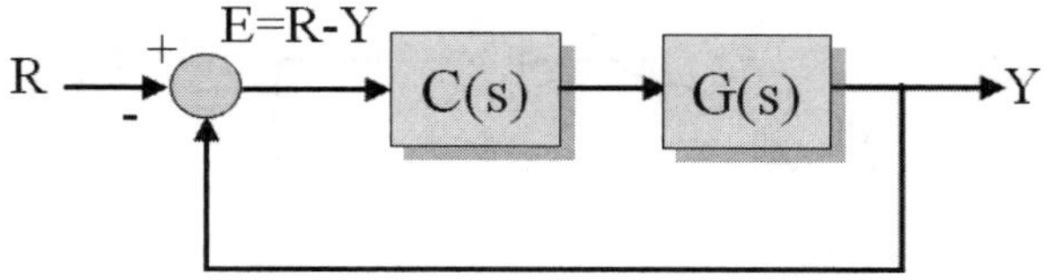

FIGURE 11.4
Block scheme for the control of the RTP system.

Let us consider the following linear model:

$$\begin{aligned} \dot{\mathbf{x}} &= \mathrm{A}\mathbf{x} + \mathrm{B}u \\ y &= \mathrm{C}\mathbf{x} \end{aligned} \tag{11.13}$$

with

$$\mathrm{A} = \begin{bmatrix} -0.0682 & 0.0149 & 0 \\ 0.0458 & -0.1181 & 0.0218 \\ 0 & 0.04683 & -0.1008 \end{bmatrix}; \mathrm{B} = \begin{bmatrix} 0.5122 \\ 0.5226 \\ 0.4185 \end{bmatrix}; \mathrm{C} = \begin{bmatrix} 0 & 1 & 0 \end{bmatrix}; \mathrm{D} = 0;$$

The state vector $\mathbf{x}$ represents the temperatures of three tungsten halogen lamps which are tied into one actuator and only the central one used for feedback. The reader is referred to the book by Franklin, Powell and Emami-Naeini for a deeper discussion on the linear model of the RTP system and its control by classical methods.
Consider now a H_∞ control problem. Let's start from the control scheme reported in Figure 11.4. The objective of the control (beyond closed-loop stability) is to track the input, i.e., to minimize the error $e(t) = r(t) - y(t)$. So, the first step is to rewrite the problem in terms of this block scheme. To do this, let's write the state-space equations by introducing the variable $\tilde{y} = r - y$ (that is the input of the controller):

$$\begin{aligned} \dot{\mathbf{x}} &= \mathrm{A}\mathbf{x} + \mathrm{B}u \\ \tilde{y} &= w - \mathrm{C}\mathbf{x} \end{aligned} \tag{11.14}$$

To satisfy the objective of input tracking, we choose $z_\infty = e$ and $w = r$. This is the worst case for the error, since for a SISO system the H_∞ norm represents the maximum value of the frequency response, or in other words the maximum amplification for any sinusoidal input.
The whole equations of the system to be controlled are rewritten as:

$$\begin{aligned} \dot{\mathbf{x}} &= \mathrm{A}\mathbf{x} + \mathrm{B}u \\ z_\infty &= w - \mathrm{C}\mathbf{x} \\ \tilde{y} &= w - \mathrm{C}\mathbf{x} \end{aligned} \tag{11.15}$$

from which it is possible to define $\mathrm{B}_1 = \begin{bmatrix} 0 & 0 & 0 \end{bmatrix}^T$, $\mathrm{B}_2 = \mathrm{B}$, $\mathrm{C}_1 = \mathrm{C}_2 = -\mathrm{C}$, $D_{11} = D_{21} = 1$, and $D_{12} = D_{22} = 0$.
The realization matrix of the process can be written as:

$$P(s) = \left[\begin{array}{ccc|cc} -0.0682 & 0.0149 & 0 & 0 & 0.5122 \\ 0.0458 & -0.1181 & 0.0218 & 0 & 0.5226 \\ 0 & 0.04683 & -0.1008 & 0 & 0.4185 \\ \hline 0 & -1 & 0 & 1 & 0 \\ 0 & -1 & 0 & 1 & 0 \end{array}\right];$$

Applying the procedure for the design of the H_∞ control by using the MATLAB®

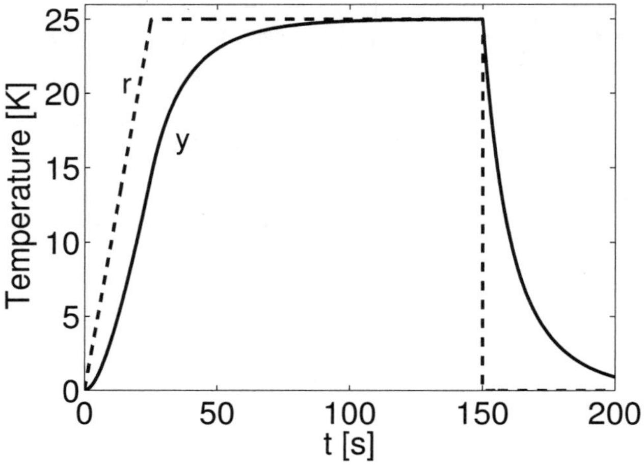

FIGURE 11.5
Response of the closed-loop system with the designed H_∞ controller $K(s)$ as in equation (11.16).

command **hinfric**, we obtain a performance index $\gamma_{opt} = 1.01$ and the following compensator:

$$K(s) = \frac{1.1285(s+0.099)(s+0.071)}{(s+1908)(s+0.14)(s+0.088)} \tag{11.16}$$

The temperature tracking response of the closed-loop system is reported in Figure 11.5 where the input signal (dashed line) is followed by the response (continuous line).

Other examples of H_∞ control are addressed in the following chapters where a very general technique is used to solve them, based on the so-called Linear Matrix Inequalities (LMI).

11.3 The H_∞ linear control and the uncertainty problem

As we saw in the introduction, uncertainty is a problem of fundamental importance in control systems. In the case of H_∞ control, we have a significant result permitting us to check the robustness of the control system to uncertainties.

Consider again the control scheme shown in Figure 11.2. As discussed above, the objective of the H_∞ control is to determine a stabilizing controller that ensures that $\|T_{zw}(s)\|_\infty < \gamma$. In fact, the transfer matrix depends on the characteristics of process $P(s)$ to be controlled and on the choice of the

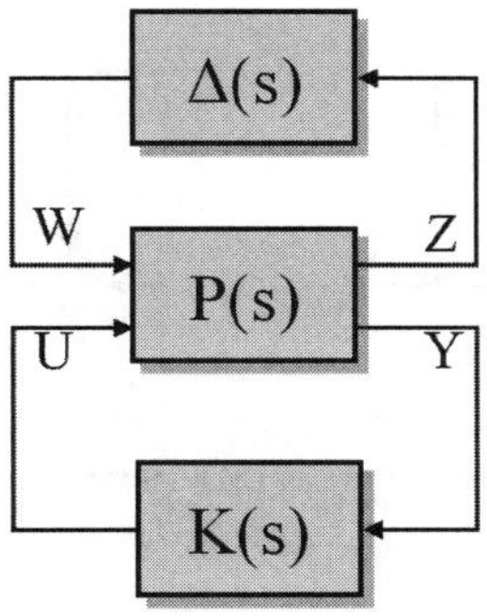

FIGURE 11.6
Scheme of robust control with uncertainty $\Delta(s)$.

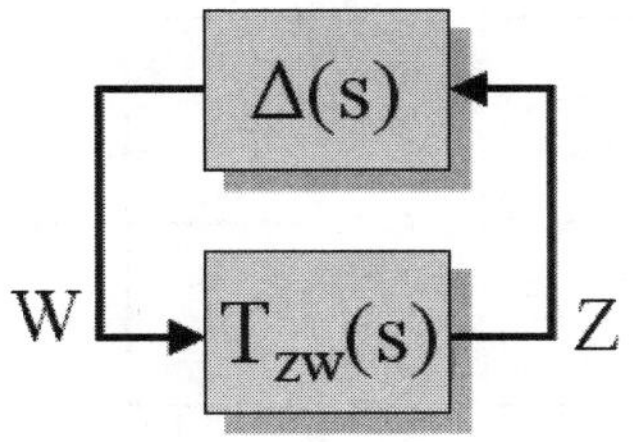

FIGURE 11.7
Feedback connection between $\Delta(s)$ and $T_{zw}(s)$.

compensator, i.e., $T_{zw} = T_{zw}(P, K)$. Thus, for this reason, one can act on K so as to obtain $\|T_{zw}(s)\|_\infty < \gamma$.

Now, let's consider the uncertainty $\Delta(s)$ defined as in Figure 11.6. The uncertainty is modeled through a transfer function from z to w. We will see below that this model can represent various types of uncertainty (multiplicative and additive, for example).

The scheme in Figure 11.6 can be simplified as in Figure 11.7, which highlights that the connection between $\Delta(s)$ and $T_{zw}(s)$ is a feedback connection. At this point, system stability can be found by applying the theorem 24. If system $\Delta(s)T_{zw}(s)$ is bounded real, then the feedback system is asymptotically stable. Since the H_∞ norm is a consistent norm, then $\|\Delta(s)T_{zw}(s)\|_\infty \leq \|\Delta(s)\|_\infty \|T_{zw}(s)\|_\infty$, so if $\|T_{zw}(s)\|_\infty < \gamma$ then for any uncertainty that $\|\Delta(s)\|_\infty < \frac{1}{\gamma}$ we are guaranteed that the feedback system is asymptotically stable.

Therefore, for the H_∞ control system, we are guaranteed it is robust to any stable perturbation having H_∞ norm less than $\frac{1}{\gamma}$.

Now let's see how additive and multiplicative uncertainty can be repre-

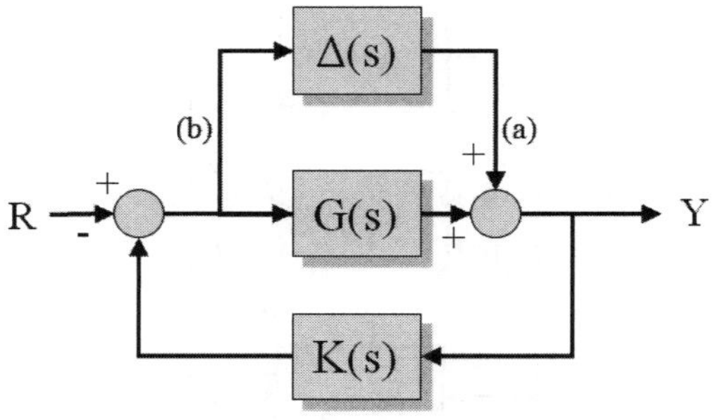

FIGURE 11.8
Scheme representing the additive uncertainty on $G(s)$.

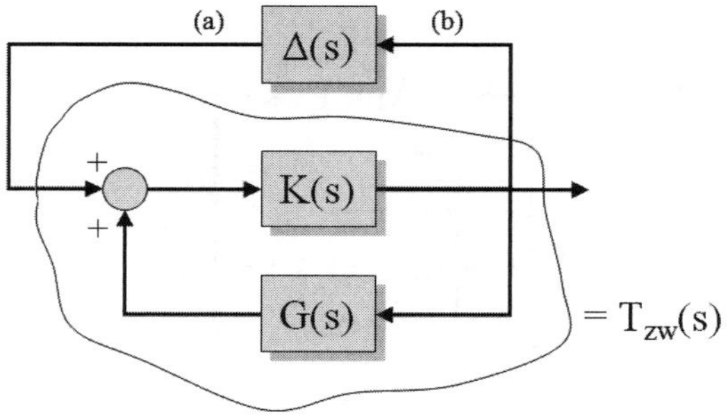

FIGURE 11.9
Formulation of additive uncertainty in terms of feedback between $\Delta(s)$ and $T_{zw}(s)$.

sented according to the scheme in Figure 11.7. First, consider the additive uncertainty as in Figure 11.8. Opening the circuit at the points marked (a) and (b), the scheme can be redrawn as in Figure 11.8, showing that by choosing $T_{zw} = K(I - GK)^{-1}$ the additive uncertainty can be modeled according to the scheme in Figure 11.7.

As for the multiplicative uncertainty of type $\tilde{G}(s) = G(s)(I + \Delta(s))$, represented by Figure 11.10, we can use an equivalent scheme as in Figure 11.11. So, multiplicative uncertainty can be modeled as in Figure 11.7 as long as you choose $T_{zw}(s) = KG(I - KG)^{-1}$.

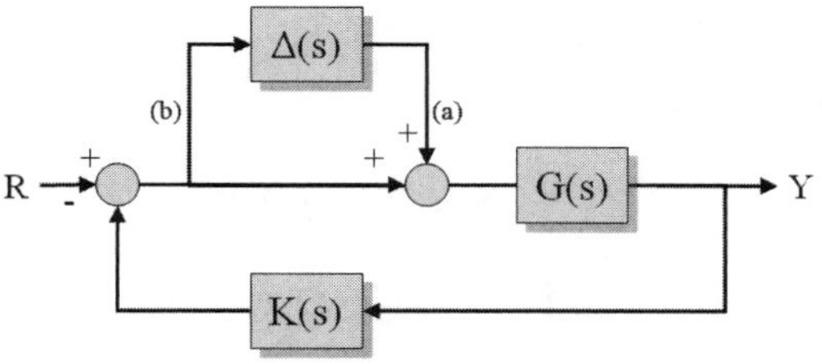

FIGURE 11.10
Scheme representing the multiplicative uncertainty on $G(s)$.

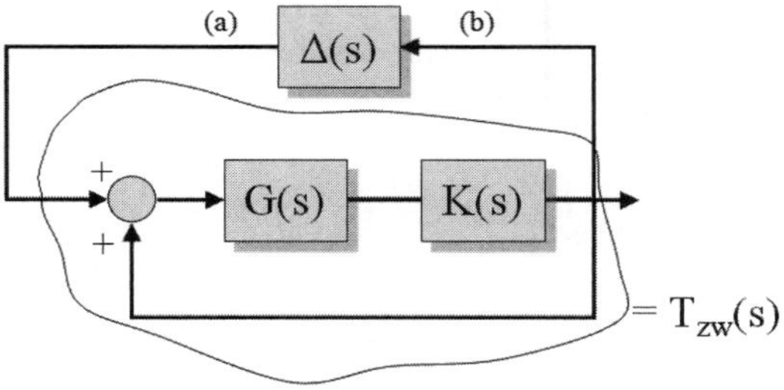

FIGURE 11.11
Formulation of multiplicative uncertainty in terms of feedback between $\Delta(s)$ and $T_{zw}(s)$.

11.4 Exercises

1. Calculate analytically the H_∞ norm for the system with transfer function $G(s) = \frac{2s+1}{s+2}$.
2. Given the continuous-time system with state-space matrices:

$$A = -1; \quad B = \begin{bmatrix} 1 & 2 \end{bmatrix}; \quad C = \begin{bmatrix} 3 \\ 4 \end{bmatrix}$$

   determine analytically the H_∞ norm.
3. Given the continuous-time system with state-space matrices:

$$\mathrm{A} = -3; \quad \mathrm{B} = \begin{bmatrix} 1 & 3 \end{bmatrix}; \quad \mathrm{C} = \begin{bmatrix} 6 \\ 9 \end{bmatrix}$$

determine analytically the H_∞ norm.

4. Given the system

$$\begin{cases} \dot{x}_1 = x_1 + u + 2w \\ y = x_1 + w \\ z = 2x_1 + 2u \end{cases}$$

calculate analytically the compensator $C(s)$ with the H_∞ control technique.

5. Calculate the performance of the H_∞ control for the system:

$$\begin{cases} \dot{x} = 2x + 0.1w + u \\ z = 2x + 0.7u \\ y = -x + 2w \end{cases} \tag{11.17}$$

12

Linear Matrix Inequalities for optimal and robust control

CONTENTS

This chapter introduces a technique based on matrix inequalities (the so-called Linear Matrix Inequalities or LMIs) which solves many control problems through a very general formulation. After expressing the main ideas on which the approach relies, the most important control problems will be reformulated according to LMI.

12.1 Definition and properties of LMI

A LMI is a relation of type:

$$A(\mathbf{x}) = A_0 + x_1 A_1 + x_2 A_2 + \ldots + x_n A_n < 0 \tag{12.1}$$

where $\mathbf{x} = \begin{pmatrix} x_1, & x_2, & \ldots, & x_n \end{pmatrix}$ is a vector of unknowns, said decisional variables or optimization variables, $A_0, A_1, A_2, \ldots, A_n$ are assigned symmetrical matrices, and where the inequality has to be seen in matricial sense ($A(\mathbf{x})$ is a negative definite matrix).

Note that constraints of type $A(\mathbf{x}) > 0$ and $A(\mathbf{x}) < B(\mathbf{x})$ can be reformulated according to (12.1), considering $-A(\mathbf{x}) < 0$ and $A(\mathbf{x}) - B(\mathbf{x}) < 0$.

Also note that all the points $\mathbf{x}$ that satisfy the LMI inequality (12.1) constitute a convex set. In fact if $A(\mathbf{y}) < 0$ and $A(\mathbf{z}) < 0$, then $A(\frac{\mathbf{y}+\mathbf{z}}{\mathbf{2}}) < 0$.

From the convexity property derives the important consequence that, although the inequality (12.1) has no analytical solution in the general case, it can be solved numerically with the guarantee to find a solution if it exists.

Moreover, since the set of solutions, also called "feasible set," is a convex subset of $\mathbb{R}^n$, finding a solution for the LMI (12.1) is a convex optimization problem. Recall that an optimization problem consists of finding a minimum (or a maximum) in certain regions defined by certain constraints on the independent variables.

Furthermore, all the conditions

$$\begin{cases} A_1(\mathbf{x}) < 0 \\ A_2(\mathbf{x}) < 0 \\ \ldots \\ A_k(\mathbf{x}) < 0 \end{cases}$$

are equivalent to one single LMI of type:

$$A(\mathbf{x}) = \begin{bmatrix} A_1(\mathbf{x}) & 0 & \ldots & 0 \\ 0 & A_2(\mathbf{x}) & \ldots & 0 \\ \vdots & & \ddots & \vdots \\ 0 & 0 & \ldots & A_k(\mathbf{x}) \end{bmatrix} < 0$$

Finally, it is important to notice that the nonlinear inequalities of convex type can be transformed in LMI. For example, the following set of inequalities:

$$\begin{cases} R(\mathbf{x}) > 0 \\ Q(\mathbf{x}) - S(\mathbf{x}) R^{-1}(\mathbf{x}) S^T(\mathbf{x}) > 0 \end{cases} \tag{12.2}$$

with $R(\mathbf{x})$ and $Q(\mathbf{x})$ symmetrical is equivalent to the LMI:

$$\begin{bmatrix} Q(\mathbf{x}) & S(\mathbf{x}) \\ S^T(\mathbf{x}) & R(\mathbf{x}) \end{bmatrix} > 0 \tag{12.3}$$

In this way, the nonlinear inequality (12.2) can be transformed in a linear inequality (12.3).

In many control problems the LMI inequalities are not in form (12.1), but are structured as follows:

$$\mathbf{L}(X_1, \ldots, X_n) < \mathbf{R}(X_1, \ldots, X_n) \tag{12.4}$$

where $\mathbf{L}$ and $\mathbf{R}$ are affine functions of some structured matricial variables $\mathrm{X}_1, \ldots, \mathrm{X}_n$.

For example, consider

$\mathrm{A}^T\mathrm{X} + \mathrm{XA} < 0$ con $\mathrm{A} = \begin{bmatrix} a_{11} & a_{12} \\ a_{21} & a_{22} \end{bmatrix}$. This LMI problem has a particular importance for the study of the stability of linear time-invariant systems. In fact, it is equivalent to the Lyapunov theorem for time-invariant systems. A linear time-invariant system is asymptotically stable if the LMI $\mathrm{A}^T\mathrm{X}+\mathrm{XA} < 0$ admits a solution $\mathrm{X} > 0$. If we consider $\mathrm{X} = \begin{bmatrix} x_{11} & x_{12} \\ x_{12} & x_{22} \end{bmatrix}$, it is immediate to verify that the inequality $\mathrm{A}^T\mathrm{X} + \mathrm{XA} < 0$ can be written as $\mathrm{A}_1 x_{11} + \mathrm{A}_2 x_{12} + \mathrm{A}_3 x_{22}$ with A_1, A_2 and A_3 appropriate matrices.

The LMIs can be applied to control, identification and design problems. Their importance for these kinds of problems is motivated by several considerations: the design specifications and the design constraints can be expressed in LMI terms; the control problems can be reformulated in terms of convex optimization problems; many of the control problems lack analytical solutions, but they can be treated in LMI terms.

Furthermore, in the period in which the LMIs were discovered, the interior point method to solve convex optimization problems was developed. The development of this optimization method was a further motivation to use the LMI.

12.2 LMI problems

There exist some convex optimization problems which can be considered canonical. The analysis of control problems often comes down to three LMI canonical problems which will be described in detail.

12.2.1 Feasibility problem

The feasibility problem consists of finding a solution $\mathbf{x}$ to the LMI problem

$$\mathrm{A}(\mathbf{x}) < 0 \tag{12.5}$$

and it is equivalent to solving the optimization problem:

$$\min t \ \texttt{with constraint} \ \mathrm{A}(\mathbf{x}) < t\mathrm{I} \tag{12.6}$$

with $t < 0$. This problem can be solved with MATLAB® command `feasp`.

12.2.2 Linear objective minimization problem

Minimizing a convex objective function with a LMI is still a convex problem. In particular, if the objective function is linear it is said to be a *linear objective minimization problem.* This problem is defined by:

$$\min \mathrm{C}^T\mathbf{x} < 0 \ \texttt{with constraint}\mathrm{A}(\mathbf{x}) < 0 \tag{12.7}$$

This problem can be solved using the MATLAB command `mincx`.

12.2.3 Generalized eigenvalue minimization problem

This problem consists in minimizing

$$\min \lambda \tag{12.8}$$

with the constraints

$$\begin{cases} \mathrm{A}(\mathbf{x}) < \lambda \mathrm{B}(\mathbf{x}) \\ \mathrm{B}(\mathbf{x}) > 0 \\ \mathrm{C}(\mathbf{x}) < 0 \end{cases} \tag{12.9}$$

This problem can be solved using the MATLAB command `gevp`.

A simplified version of the problem is given by the objective to minimize

$$\min \lambda \tag{12.10}$$

with the constraints

$$\begin{cases} \lambda \mathrm{I} - \mathrm{A}(\mathbf{x}) > 0 \\ \mathrm{B}(\mathbf{x}) > 0 \end{cases} \tag{12.11}$$

Solving a LMI problem means determining firstly if the problem admits solutions, and so calculating the feasible solution.

The interior point optimization algorithms developed by Nesterov and Nemirovski allowed to solve in an efficient and fast way the previously discussed canonical generical LMI problems.

12.3 Formulation of control problems in LMI terms

Many control problems can be formulated in terms of LMI inequalities. In this section we will examine some important examples.

12.3.1 Stability

The state $\mathbf{x} = 0$ of the system $\dot{\mathbf{x}} = \mathrm{A}\mathbf{x}$ is asymptotically stable if there exists a matrix $\mathrm{P} > 0$ which satisfies $\mathrm{A}^T\mathrm{P} + \mathrm{PA} < 0$. The LMI problem is feasible if the system is asymptotically stable. To study the stability of a linear time-invariant system through LMI, we have to solve a feasibility problem defined by:

$$\begin{aligned} &\mathrm{P} > 0 \\ &\mathrm{A}^T\mathrm{P} + \mathrm{PA} < 0 \end{aligned} \tag{12.12}$$

This LMI problem admits a closed form solution, given by the Lyapunov theorem for linear time-invariant systems.

12.3.2 Simultaneous stabilizability

The LMI problem (12.12) can be adapted properly for the study of closed-loop stability. In this case the feasibility problem to study is

$$\begin{aligned} &\mathrm{P} > 0 \\ &\mathrm{A}_c\mathrm{P} + \mathrm{PA}_c^T < 0 \end{aligned} \tag{12.13}$$

with $\mathrm{A}_c = \mathrm{A} - \mathrm{BK}$ closed-loop state matrix. Strictly speaking note that we considered the equivalent problem of studying the stability of system A_c^T.

Substituting the expression of A_c in the equation (12.13), we obtain

$$\begin{aligned} &\mathrm{P} > 0 \\ &(\mathrm{A} - \mathrm{BK})\mathrm{P} + \mathrm{P}(\mathrm{A}^T - \mathrm{K}^T\mathrm{B}^T) < 0 \end{aligned} \tag{12.14}$$

and putting $\mathrm{Q} = \mathrm{KP}$ we obtain

$$\begin{aligned} &\mathrm{P} > 0 \\ &\mathrm{AP} - \mathrm{BQ} + \mathrm{PA}^T - \mathrm{Q}^T\mathrm{B}^T < 0 \end{aligned} \tag{12.15}$$

The LMI problem (12.15) in the two unknowns P and Q permits us to find the control law which stabilizes the system under the assumption that it is controllable. Since P is positive definite, we can calculate the inverse and get K: $\mathrm{K} = \mathrm{QP}^{-1}$.

The peculiarity of this method is that it can be applied to the *simultaneous stabilizability* problem. This term indicates the possibility to find a control law which stabilizes simultaneously two or more systems.

So, consider m systems (all of order n) $S_1(\mathrm{A}_1, \mathrm{B}_1)$, ..., $S_m(\mathrm{A}_m, \mathrm{B}_m)$. The problem of the existence of a control law $\mathbf{u} = -\mathrm{K}\mathbf{x}$ that stabilizes simultaneously these systems can be viewed as a feasibility problem defined by the following LMI inequalities:

$$\begin{aligned} &\mathrm{P} > 0 \\ &\mathrm{A}_1\mathrm{P} - \mathrm{B}_1\mathrm{Q} + \mathrm{PA}_1^T - \mathrm{Q}^T\mathrm{B}_1^T < 0 \\ &\dots \\ &\mathrm{A}_m\mathrm{P} - \mathrm{B}_m\mathrm{Q} + \mathrm{PA}_m^T - \mathrm{Q}^T\mathrm{B}_m^T < 0 \end{aligned} \tag{12.16}$$

We will see in the following chapter that there exist some analytical conditions for the study of simultaneous stabilizability of two systems, while for more systems the numerical approach here described has to be used. In paragraph 12.5 an example of simultaneous stabilizability is described.

12.3.3 Positive real lemma

The problem of stating if a system is passive $\int_0^T u^T(t)y(t)dt \geq 0$ can be solved through the positive real lemma discussed in Chapter 10. This problem can also be formulated in LMI terms. Also in this case we are talking about a feasibility problem. Particularly, the positive real lemma can be formulated in terms of two LMIs defined by

$$\begin{array}{l} \mathrm{P} > 0 \\ \left[\begin{array}{cc} \mathrm{A}^T\mathrm{P} + \mathrm{PA} & \mathrm{PB} - \mathrm{C}^T \\ \mathrm{B}^T\mathrm{P} - \mathrm{C} & -\mathrm{D}^T - \mathrm{D} \end{array}\right] \leq 0 \end{array} \tag{12.17}$$

The problem is feasibile if the assumptions of the positive real lemma are verified.

12.3.4 Bounded real lemma

Remember that a system is called bounded real if for any input $u(t)$ the output is such that $\int_0^T y^T(t)y(t)dt \leq \int_0^T u^T(t)u(t)dt$. For linear time-invariant systems this property can be verified through the bounded real lemma, or in LMI terms verifying that the problem is defined by

$$\begin{array}{l} \mathrm{P} > 0 \\ \left[\begin{array}{cc} \mathrm{A}^T\mathrm{P} + \mathrm{PA} + \mathrm{C}^T\mathrm{C} & \mathrm{PB} + \mathrm{C}^T\mathrm{D} \\ \mathrm{B}^T\mathrm{P} + \mathrm{D}^T\mathrm{C} & \mathrm{D}^T\mathrm{D} - \mathrm{I} \end{array}\right] \leq 0 \end{array} \tag{12.18}$$

is feasible.

12.3.5 Calculating the H_∞ norm through LMI

The H_∞ norm of a system $\mathrm{G}(s) = \mathrm{C}(s\mathrm{I} - \mathrm{A})^{-1}\mathrm{B}$ is α if $\|\mathrm{G}(j\omega)\|_S \leq \alpha \ \forall\omega$ or equivalently $\|\alpha^{-1}\mathrm{G}(j\omega)\|_S \leq 1$, i.e., if the system $\tilde{\mathrm{G}}(j\omega) = \alpha^{-1}\mathrm{G}(j\omega)$ is bounded real. The approach shown in Chapter 11 is based on the construction of a Hamiltonian matrix $\mathrm{H} = \left[\begin{array}{cc} \mathrm{A} & \frac{\mathrm{BB}^T}{\alpha^2} \\ -\mathrm{C}^T\mathrm{C} & -\mathrm{A}^T \end{array}\right]$ and on the verification of the minimum value of α for which the matrix has no eigenvalues on the imaginary axis.

Considering that this problem is associated to the Riccati equation $\mathrm{A}^T\mathrm{P} + \mathrm{PA} + \mathrm{C}^T\mathrm{C} + (\alpha^2)^{-1}\mathrm{PB}^T\mathrm{BP} = 0$, the problem can be reformulated with a matrix inequality $\mathrm{A}^T\mathrm{P} + \mathrm{PA} + \mathrm{C}^T\mathrm{C} + (\alpha^2)^{-1}\mathrm{PB}^T\mathrm{BP} < 0$, in which as you can see a nonlinear term in P appears. Again, this inequality can be reformulated as a linear matrix inequality:

$$\begin{array}{l} \mathrm{P} > 0 \\ \begin{bmatrix} \mathrm{A}^T\mathrm{P} + \mathrm{PA} + \mathrm{C}^T\mathrm{C} & \mathrm{PB} \\ \mathrm{B}^T\mathrm{P} & 0 \end{bmatrix} \leq \alpha^2 \begin{bmatrix} \varepsilon \mathrm{I} & 0 \\ 0 & \mathrm{I} \end{bmatrix} \end{array} \tag{12.19}$$

The LMI problem (12.19) is a generalized eigenvalue problem of type $F(x) < \alpha^2 E(x)$. ε is a small quantity introduced to obtain a numerical solution of the problem.

12.4 Solving a LMI problem

In MATLAB® it is possible to define any LMI problem and so solve it with a command `feasp`, `mincx` or `gevp`. The definition of the LMI problem does not depend on the type of problem we want to solve and it is completely general.

The definition of a LMI problem begins with the command `setlmis` and ends with the command `getlmis`. At first it is necessary to define the decisional variables, i.e., the unknowns, of the LMI problem. The command to use is `lmivar` with the following syntax:

`P=lmivar(type,structure)`

This command permits us to choose unknown symmetrical matrices, rectangular matrices or matrices of other type. Depending on the chosen matrix type the structure contains different information:

- if type=1, matrix P is square symmetrical and the structure element (i,1) specifies the dimension of the i-block, while the structure element(i,2) specifies the type of block (0 for scalar blocks of type $x * \mathrm{I}$, 1 for complete blocks, -1 for zero-blocks);
- if type=2, matrix P is rectangular of size $m \times n$ as specified in structure=[m,n];
- if type=3, matrix P is of other type.

MATLAB® exercise 12.1

Define the matrices of decisional variables X_1 symmetrical 3×3, X_2 rectangular 2×4 and $\mathrm{X}_3 = \begin{bmatrix} \Delta & 0 & 0 \\ 0 & \delta_1 & 0 \\ 0 & 0 & \delta_2 \mathrm{I}_2 \end{bmatrix}$ with I_2 identity matrix of order two and Δ 5×5 matrix:

```
X1=lmivar(1,[3 1]);
X2=lmivar(2,[2,4]);
X3=lmivar(1,[5 1; 1 0; 2 0]);
```

A LMI inequality can be specified in MATLAB defining each of its constituent terms with command `lmiterm`. The syntax of the command is:

`lmiterm(termID, A,B,flag)`

In the termID parameter different information are summarized. The first element answers the question: at which LMI belongs the term? termID(1) in fact can be $+n$ or $-n$ depending on whether the specified term appears, respectively, in the member in the left-hand term or in the right-hand term of the $n-$th matrix inequality.

The second and the third element answer the question: at which block belongs the specified term? In fact, termID(2:3) can be equal to [00] for external factors or to $[i, j]$ to indicate the ij block of a generic matrix term that appears in the LMI.

Finally, the fourth element permits us to specify which type of term we are adding to the LMI problem. In fact, termID(4) can be 0, X or -X depending on whether the term is constant, of type AXB or $AX'B$.

The other parameters of the command `lmiterm` are more intuitive. A and B represent the matrices that left- or right- multiply the variable respectively, while the flag if set on 's' permits us to specify with a single command that in the LMI appears not only the given term, but also its symmetrical.

In MATLAB it is also possible to use a graphical interface to define a LMI problem. The interface can be called through the command `lmiedit` and permits us to obtain the sequence of the commands that define the LMI problem.

MATLAB® exercise 12.2

As example, we show the definition of a LMI problem for the study of the stability of a linear time-invariant system with state matrix A $= diag\{-1, -2, \dots, -10\}$.

```
A=diag([-1:-1:-10]);
setlmis([]);
P=lmivar(1,[10 1]);
lmiterm([-1 1 1 P],1,1);           % LMI #1: P
lmiterm([2 1 1 P],A',1,'s');       % LMI #2: A'*P+P*A
stabproblem=getlmis;
```

Once defined, it is possible to solve the LMI problem with the command `feasp`:

```
[tmin,Psol] = feasp(stabproblem);
```

As you can see we obtain a value of $t < 0$. To obtain the solution matrix of the problem we must pass from the decisional matrices to the correspondent matrix with the command `dec2mat`:

```
Pmatrice=dec2mat(stabproblem,Psol,P);
```

MATLAB® exercise 12.3

Consider another example, calculating the H_∞ norm for the system A $=$ $diag\{-1, -2, \dots, -10\}$; B $= \begin{bmatrix} 1 \\ 1 \\ \vdots \\ 1 \end{bmatrix}$; C $= \begin{bmatrix} 1 & 1 & \dots & 1 \end{bmatrix}$.

We observe that the system is a relaxation system and it is also a strictly proper system.

This implies that the H_∞ norm of the system coincides with the value of the frequency response obtained for $\omega \to 0$ so it can be calculated considering $G(0) = \mathrm{C}(-\mathrm{A})^{-1}\mathrm{B}$.
Let us begin defining the system with MATLAB:

```
>> A=diag([-1 -2 -3 -4 -5 -6 -7 -8 -9 -10]);
>> B=ones(10,1);
>> C=ones(1,10);
>> system=ss(A,B,C,0);
```

Then calculate the H_∞ norm exploiting the peculiarities of the system as $\|G(s)\|_\infty = \mathrm{C}(-\mathrm{A})^{-1}\mathrm{B}$:

```
>> normmethod1=C*inv(-A)*B
```

A second method to calculate the H_∞ norm is based on the use of the command `normhinf`:

```
>> normmethod2=normhinf(system)
```

In the LMI toolbox a specific command to calculate the H_∞ norm also exists. In this case the system has to be defined as object `ltisys` and the norm can be calculated as follows:

```
>> g=ltisys(A,B,C,0)
>> [normmethod3, pekf]=norminf(g)
```

Finally, we show the calculation of the H_∞ norm through the definition of a LMI problem which makes use of the equation (12.19):

```
>> setlmis([]);
>> P=lmivar(1,[10 1]);
>> lmiterm([-1 1 1 P],1,1);      % LMI #1: P>0
>> lmiterm([2 1 1 P],A',1,'s');  % LMI #2: A'*P+P*A
>> lmiterm([2 1 1 0],C'*C);      % LMI #2: C'*C
>> lmiterm([2 2 1 P],B',1);      % LMI #2: B'*P
>> lmiterm([-2 1 1 0],0.00001);  % LMI #2: epsilon
>> lmiterm([-2 2 2 0],1);        % LMI #2: 1
>> LMIproblem=getlmis;
>> [a1,po]=gevp(LMIproblem,1);
>> normmethod4=sqrt(a1)
```

Note that the command `gevp` requires that the LMIs in which the generalized eigenvalue λ to minimize appears are written at the end of the list of LMIs. In addition we have to specify in how many LMIs appears λ (in this case it appears in a single LMI).

12.5 LMI problem for simultaneous stabilizability

The problem of simultaneous stability is illustrated through an example. Then, an application to the control of nonlinear circuits will be discussed.

Consider the system

$$\begin{cases} \dot{x}_1 = x_1^2 + 3x_2 + 2u \\ \dot{x}_2 = x_1 + x_2 \end{cases} \tag{12.20}$$

At $u = 1$ the system admits two equilibrium points $(\bar{x}_1, \bar{x}_2) = (1, -1)$ and $(\tilde{x}_1, \tilde{x}_2) = (2, -2)$. Suppose we want to stabilize the system around both equilibrium points with a unique control law.

The linearized system around the two equilibrium is described by the state matrices:

$$\mathrm{A}_1 = \begin{bmatrix} 2 & 3 \\ 1 & 1 \end{bmatrix}; \mathrm{A}_2 = \begin{bmatrix} 4 & 3 \\ 1 & 1 \end{bmatrix}; \mathrm{B} = \mathrm{B}_1 = \mathrm{B}_2 = \begin{bmatrix} 2 \\ 0 \end{bmatrix}$$

Therefore it is necessary to solve the LMI problem:

$$\begin{array}{l} \mathrm{P} > 0 \\ \mathrm{A}_1\mathrm{P} - \mathrm{BQ} + \mathrm{PA}_1^T - \mathrm{Q}^T\mathrm{B}^T < 0 \\ \mathrm{A}_2\mathrm{P} - \mathrm{BQ} + \mathrm{PA}_2^T - \mathrm{Q}^T\mathrm{B}^T < 0 \end{array} \tag{12.21}$$

Once the matrices are defined in MATLAB®, the LMI problem can be defined through the commands:

```
>> setlmis([]);
>> P=lmivar(1,[2 1]);
>> Q=lmivar(2,[1,2]);
>> lmiterm([-1 1 1 P],1,1);
>> lmiterm([2 1 1 P],A1,1,'s');
>> lmiterm([2 1 1 Q],B,-1,'s');
>> lmiterm([3 1 1 P],A2,1,'s');
>> lmiterm([3 1 1 Q],B,-1,'s');
>> stabilz2=getlmis;
```

The solution can be calculated through the commands:

```
>> [tmin,xfeas] = feasp(stabilz2);
>> Pvalue=dec2mat(stabilz2,xfeas,P);
>> Qvalue=dec2mat(stabilz2,xfeas,Q);
>> k=Qvalue*inv(Pvalue)
```

Note that the obtained value guarantees that the closed-loop eigenvalues have negative real part whether the system works around the equilibrium point $\bar{\mathbf{x}}$ or around $\tilde{\mathbf{x}}$:

```
>> eigenvaluessystem1=eig(A1-B*k)
>> eigenvaluessystem2=eig(A2-B*k)
```

As crosscheck, verify that the obtained controller for example considering $\lambda_1 = -1$ and $\lambda_2 = -0.5$ in the closed-loop system $\mathrm{A}_1 - \mathrm{BK}$ does not guarantee the stability of $\mathrm{A}_2 - \mathrm{BK}$. To do this we use the commands:

```
>> openloopeigenvalues=eig(A1)
>> K=acker(A1,B,[-1 -0.5])
>> eig(A1-B*K)
>> eigenvaluessystem1=eig(A1-B*K)
>> eigenvaluessystem2=eig(A2-B*K)
```

MATLAB® exercise 12.4

In this MATLAB exercise, the LMI approach to simultaneous stability is adopted for the design of an asymptotic observer for a nonlinear circuit.

Let us first briefly discuss the observer design based on LMI for a system described by:

$$\dot{\mathbf{X}} = \mathrm{A}\mathbf{X}, \tag{12.22}$$

The dynamical equations of the observer system are:

$$\dot{\hat{\mathbf{X}}} = \mathrm{A}\hat{\mathbf{X}} + \mathrm{K}e, \tag{12.23}$$

where K are the observer gains and the error is defined as $e = \mathrm{C}\mathbf{X} - \mathrm{C}\hat{\mathbf{X}}$ with (A, C) being the state matrices of the system. K has to be chosen in order to ensure the stability of the error system. This can be done following a LMI approach defined starting from the following Lyapunov equation:

$$\mathrm{A}_\mathrm{O}^\mathrm{T}\mathrm{P} + \mathrm{P}\mathrm{A}_\mathrm{O} = -\bar{\mathrm{Q}}, \tag{12.24}$$

with P and $\bar{\mathrm{Q}}$ positive definite matrices and $\mathrm{A}_\mathrm{O} = \mathrm{A} - \mathrm{KC}$ the state matrix of the error system.
The corresponding LMI problem will be the following:

$$\begin{cases} \mathrm{A}_\mathrm{O}^\mathrm{T}\mathrm{P} + \mathrm{P}\mathrm{A}_\mathrm{O} < 0 \\ \mathrm{P} > 0 \end{cases} \tag{12.25}$$

where the first constraint is:

$$(\mathrm{A} - \mathrm{KC})^\mathrm{T}\mathrm{P} + \mathrm{P}(\mathrm{A} - \mathrm{KC}) < 0 \tag{12.26}$$

Define now $\mathrm{Q} = \mathrm{PK}$ and substitute in Eq. (12.25):

$$\begin{cases} \mathrm{A}^\mathrm{T}\mathrm{P} - \mathrm{C}^\mathrm{T}\mathrm{Q}^\mathrm{T} + \mathrm{PA} - \mathrm{QC} < 0 \\ \mathrm{P} > 0 \end{cases} \tag{12.27}$$

Once P and Q solving problem (12.27) are obtained, K is calculated as $\mathrm{K} = \mathrm{P}^{-1}\mathrm{Q}$. The problem is feasible if system (A, C) is observable.
Let us now design the observer for the nonlinear circuit described by the following equations:

$$\begin{cases} \dot{x} = \alpha[y - h(x)] \\ \dot{y} = x - y + z \\ \dot{z} = -\beta y \end{cases} \tag{12.28}$$

with $h(x) = m_1 x + \frac{1}{2}(m_0 - m_1)(|x+1| - |x-1|)$ representing the piece-wise linear nonlinearity. This nonlinear system represents the dimensionless equations governing the so-called Chua's circuit, a nonlinear circuit able to show complex behavior, like chaos. The design of an observer for such kinds of systems can be considered as a solution of the nontrivial problem of chaos synchronization. In the following the parameter values are considered as: $\alpha = 9$, $\beta = 14.286$, $m_0 = -2/7$ and $m_1 = 1/7$.
In this case, the observer has to be effective for each of the possible linear regions in which the observed system and the observer may work. The advantages of the LMI approach for observer design is that other inequalities may be added to the problem so that a set of LMIs has to be solved to find the gains able to simultaneously stabilize more than one system.
Define $\mathbf{e_X} = \mathbf{X} - \hat{\mathbf{X}}$ as the state estimation error. In general, the equation that describes the error system dynamics is:

$$\dot{\mathbf{e}}_\mathbf{X} = \mathrm{A_i}\mathbf{X} - \mathrm{A_j}\hat{\mathbf{X}} - \mathrm{KC}(\mathbf{X} - \hat{\mathbf{X}}) \tag{12.29}$$

where $\mathrm{A_i}$ and $\mathrm{A_j}$ represent respectively the state matrices of the linearized systems in the i-th or j-th region of a generic PWL nonlinearity. If the systems are working in different regions of the PWL nonlinearity, the two matrices $\mathrm{A_i}$ and $\mathrm{A_j}$ are different. Otherwise (i.e., when the observer works in the same region of the observed system), the matrices $\mathrm{A_i}$ and $\mathrm{A_j}$ are equal and the error system dynamic reduces to:

$$\dot{\mathbf{e}}_\mathbf{X} = (\mathrm{A_i} - \mathrm{KC})\mathbf{e_X} \tag{12.30}$$

In this situation the observer can be designed to be stable by solving the following LMI problem:

$$\begin{cases} A_i^T P - C^T Q^T + P A_i - Q C < 0; & i = 1, \dots, q \\ P > 0 \end{cases} \tag{12.31}$$

where q is the number of regions of the considered PWL nonlinearity. This means that all the LMIs described in each region of the nonlinearity for $A_i = A_j$ have to be solved. If the LMI problem is feasible, the gain vector K able to stabilize all the possible error dynamics can be derived.

Although the error system is imposed to be stable only if the observed system and the observer are in the same PWL region (when the two systems are in different regions, the error dynamics is given by eqs. (12.29)), numerical simulations reveal that these conditions suffice to allow the design of an observer able to reconstruct the dynamics of the observed system.

The considered PWL function has three linear regions, thus the linearized system can be described by the following state matrices:

$$A_1 = A_3 = \begin{bmatrix} -\alpha m_1 & 0 & 0 \\ 1 & -1 & 1 \\ 0 & -\beta & 0 \end{bmatrix}; A_2 = \begin{bmatrix} -\alpha m_0 & 0 & 0 \\ 1 & -1 & 1 \\ 0 & -\beta & 0 \end{bmatrix}$$

$$C = C_1 = C_2 = \begin{bmatrix} 1 & 0 & 0 \end{bmatrix}$$

Once the matrices in MATLAB® are defined, the LMI problem for the observer design can be defined through the commands:

```
>> setlmis([]);
>> P=lmivar(1,[3 1]);
>> Q=lmivar(2,[1,3]);
>> lmiterm([-1 1 1 P],1,1);
>> lmiterm([2 1 1 P],A1',1,'s');
>> lmiterm([2 1 1 Q],C',-1,'s');
>> lmiterm([3 1 1 P],A2',1,'s');
>> lmiterm([3 1 1 Q],C',-1,'s');
>> stabilz2=getlmis;
```

The solution can be calculated through the commands:

```
>> [tmin,xfeas] = feasp(stabilz2);
>> Pvalue=dec2mat(stabilz2,xfeas,P);
>> Qvalue=dec2mat(stabilz2,xfeas,Q);
>> k=inv(Pvalue)*Qvalue
```

Note that the obtained value guarantees that the eigenvalues of the error system have negative real part whether the system works in the first or in the second linear region of the PWL:

```
>> eigenvaluessystem1=eig(A1-k*C)
>> eigenvaluessystem2=eig(A2-k*C)
```

The effectiveness of the designed observer is proven in Figure 12.1 where the state variables of the observer, i.e., x_O, y_O and z_O, are reported as a function of the corresponding state variables of the observed system.

12.6 Solving algebraic Riccati equations through LMI

In this paragraph two examples of control problems formulated in terms of Riccati equations and solved through LMI will be shown.

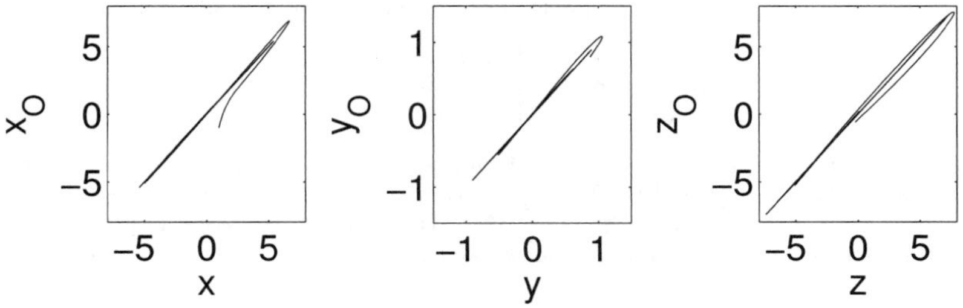

FIGURE 12.1
Representation of the observer state variables as a function of the observed system state variables: all the three state variables are correctly observed.

MATLAB® exercise 12.5

Consider first the problem of understanding when a system is bounded real or not. In this case, suppose that the system is proper, we should establish if there exists or not a solution to the Riccati equation:

$$\mathrm{A}^T\mathrm{P} + \mathrm{PA} + \mathrm{PBB}^T\mathrm{P} + \mathrm{C}^T\mathrm{C} = 0$$

Consider the system defined by

```
A=[0 1 0; 0 0 1; -5 -4 -3];
>> B=[0 0 1]';
>> C=[1 0 0];
>> system=ss(A,B,C,0);
```

The system is controllable and observable and has transfer function $G(s) = \frac{1}{s^3+3s^2+4s+5}$. It is also a stable system. It is possible to verify immediately that the system is bounded real from the Bode diagram (in fact the maximum value of $|G(j\omega)|$ is less than 1).

Define matrix Q as follows:

```
>> Q=C'*C;
```

the solution of the Riccati equation can be obtained through the command

```
>> Xare=are(A,-B*B',Q)
```

Otherwise, we can solve the LMI problem defined by the minimization of the trace of matrix P subject to a constraint of type

$$\begin{bmatrix} \mathrm{A}^T\mathrm{P} + \mathrm{PA} + \mathrm{C}^T\mathrm{C} & \mathrm{BP} \\ \mathrm{B}^T\mathrm{P} & -\mathrm{I} \end{bmatrix} \leq 0$$

The problem can be defined in MATLAB® with the following instructions (note that the problem is a problem of type `mincx`)

```
>> setlmis([]);
>> X=lmivar(1,[3 1]);
>> lmiterm([1 1 1 X],A',1,'s');
>> lmiterm([1 1 1 0],Q);
>> lmiterm([1 2 1 X],B',1);
>> lmiterm([1 2 2 0],-1);
>> lmisys=getlmis;
>> c=mat2dec(lmisys,eye(3))
>> [copt,xopt]=mincx(lmisys,c,[1e-5, 0, 0, 0, 0])
>> Xopt=dec2mat(lmisys,xopt,X)
```

We obtain the same solution that is obtained solving the Riccati equation

$$\text{Xare} = \text{Xopt} = \begin{bmatrix} 1.0528 & 0.5209 & 0.1010 \\ 0.5209 & 0.4682 & 0.1324 \\ 0.1010 & 0.1324 & 0.0445 \end{bmatrix}$$

The existence of a positive definite solution assures us that the system is bounded real. Instead consider $C = \begin{bmatrix} 10 & 0 & 0 \end{bmatrix}$. In this case the system is not bounded real. In fact, as it is possible to verify in MATLAB®, the Riccati equation does not admit solution and the LMI problem is not feasible.

MATLAB® exercise 12.6

Now consider the problem of determining the optimal controller, solving the CARE equation:

$$A^T P + PA - PBB^T P + C^T C = 0$$

To solve the problem using the approach based on the solution of the Riccati equation, we proceed in a completely analogous way to the previous case. Define the system:

```
>> A=[1 -2 1; 3 0 1; 1 -2 -1];
>> B=[1 0 1]';
>> C=[1 1 0];
>> system=ss(A,B,C,0);
```

and solve the Riccati equation with the command `are` or directly with the command `care`:

```
>> Xare=are(A,B*B',C'*C)
```

The control problem can be also solved with LMI. In this case the presence of the minus sign in the nonlinear term requires the use of some stratagems to solve the LMI problem (in fact, this term leads to a non-convex constraint). So we consider $\bar{P} = P^{-1}$, for which the Riccati equation becomes analogous to the case in the MATLAB® exercise 12.5:

$$A\bar{P} + \bar{P}A^T + \bar{P}C^T C\bar{P} - BB^T = 0$$

Doing so, we get a Riccati equation with a positive nonlinear term. At this point we can consider a new problem of type `mincx`, maximizing, rather than minimizing, the trace of matrix $\bar{P}$ (since it is the inverse of matrix P). We can obtain this changing the sign of the identity matrix with the command `c=mat2dec(lmisys,-eye(3))`, or minimizing an objective function with an opposite sign than the one in the MATLAB® exercise 12.5. The commands to define and solve the LMI problem are the following:

```
>> setlmis([]);
>> X=lmivar(1,[3 1]);
>> lmiterm([1 1 1 X],A,1,'s');
>> lmiterm([1 1 1 0],-B*B');
>> lmiterm([1 2 1 X],C,1);
>> lmiterm([1 2 2 0],-1);
>> lmisys=getlmis;
>> c=mat2dec(lmisys,-eye(3))
>> [copt,xopt]=mincx(lmisys,c,[1e-5, 0, 0, 0, 0])
>> Xopt=inv(dec2mat(lmisys,xopt,X))
```

Obviously we obtain the same solution with the two methods:

$$\text{Xare} = \text{Xopt} = \begin{bmatrix} 2.2166 & 0.2635 & 0.6757 \\ 0.2635 & 1.7438 & -0.0275 \\ 0.6757 & -0.0275 & 0.2342 \end{bmatrix}$$

from which we can immediately calculate the value of the gain K of the optimal control law: $K = B^T P = \begin{bmatrix} 2.89 & 0.24 & 0.91 \end{bmatrix}$.

12.7 Computation of gramians through LMI

We discuss here an example of LMI-based calculation of the controllability gramian with two different methods. The procedure can be easily adapted to the observability gramian.

Let us first introduce the LMI problem defined by the maximization of the trace of P subjected to the following constraints:

$$\begin{array}{l} \mathrm{P} > 0 \\ \begin{bmatrix} \mathrm{A}^T\mathrm{P} + \mathrm{PA} & \mathrm{PB} \\ \mathrm{B}^T\mathrm{P} & -\mathrm{I} \end{bmatrix} \leq 0 \end{array} \tag{12.32}$$

The solution P represents the inverse of the controllability gramian of system $S(\mathrm{A}, \mathrm{B})$.

MATLAB® exercise 12.7

Consider the system with state-space matrices:

$$\mathrm{A}_1 = \begin{bmatrix} -1 & 0 & 0 & 0 & 0 \\ 0 & -2 & 0 & 0 & 0 \\ 0 & 0 & -3 & 0 & 0 \\ 0 & 0 & 0 & -4 & 0 \\ 0 & 0 & 0 & 0 & -5 \end{bmatrix} ; \mathrm{B} = \begin{bmatrix} 1 \\ 1 \\ 1 \\ 1 \\ 1 \end{bmatrix} \tag{12.33}$$

Once the matrices in MATLAB® are defined, the linear objective problem can be defined as:

```
>> setlmis([]);
>> P=lmivar(1,[5 1]);
>> lmiterm([-1 1 1 P],1,1);
>> lmiterm([2 1 1 P],A',1,'s');
>> lmiterm([2 2 1 P],B',1);
>> lmiterm([2 2 2 0],1);
>> gramlmi1=getlmis;
```

The linear objective can be defined through the command:

```
>> c=mat2dec(gramlmi1,-eye(5));
```

note that the sign allows to maximize the trace of P. The solution can be calculated through the commands:

```
>> [tmin,xfeas] = mincx(gramlmi1,c,[1e-5 0 0 0 0]);
>> Pvalue=dec2mat(gramlmi1,xfeas,P);
>> Psol=inv(Pvalue)
```

Another approach for the calculation of the controllability gramian is based on the fact that the constraints of the LMI problem of equation (12.32) are equivalent to:

$$\begin{array}{l} \mathrm{Q} > 0 \\ \mathrm{AQ} + \mathrm{QA}^T + \mathrm{BB}^T \leq 0 \end{array} \tag{12.34}$$

with $\mathrm{Q} = \mathrm{P}^{-1}$. In this case the trace of Q has to be minimized.

MATLAB® exercise 12.8

Consider again system (12.33) and define the LMI problem:

```
>> setlmis([]);
>> P=lmivar(1,[5 1]);
>> lmiterm([-1 1 1 P],1,1);
>> lmiterm([2 1 1 P],A,1,'s');
>> lmiterm([2 1 1 0],B*B');
>> gramlmi=getlmis;
```

The linear objective is defined through the command:

```
>> c=mat2dec(gramlmi,eye(5));
```

and the solution can be calculated through the commands:

```
>> [tmin,xfeas] = mincx(gramlmi,c,[1e-5 0 0 0 0]);
>> Pvalue=dec2mat(gramlmi,xfeas,P);
```

The solution obtained is comparable with the controllability gramian calculated through the corresponding Lyapunov equation.

12.8 Computation of the Hankel norm through LMI

The Hankel norm of a system with stable transfer matrix $\mathrm{G}(s) = \mathrm{C}(s\mathrm{I} - \mathrm{A})^{-1} + \mathrm{D}$ is defined as:

$$\|\mathrm{G}(s)\|_H = \sup_{u \in L_2(\infty,0]} \left(\frac{\int_0^\infty y^T(t)y(t)dt}{\int_{-\infty}^0 u^T(t)u(t)dt} \right)^{\frac{1}{2}} \tag{12.35}$$

where

$$y(t) = \int_{-\infty}^0 \mathrm{C}e^{\mathrm{A}(t-\tau)}\mathrm{B}u(\tau)d\tau$$

and

$$L_2(\infty, 0] = \left\{ u : \left(\int_{-\infty}^0 u^T(t)u(t)dt \right)^{\frac{1}{2}} < \infty \right\}$$

It expresses the quantity of energy that can be transferred through the system from past inputs into future outputs. It can be demonstrated that:

$$\|\mathrm{G}(s)\|_H = \sigma_1 \tag{12.36}$$

where σ_1 is the maximum singular value of the system. For a LTI system the Hankel norm is also the minimum value of γ satisfying the following LMI problem:

$$\begin{cases} \mathrm{P} > 0 \\ \mathrm{Q} > 0 \\ \mathrm{A}^T\mathrm{Q} + \mathrm{QA} + \mathrm{C}^T\mathrm{C} \leq 0 \\ \begin{bmatrix} \mathrm{A}^T\mathrm{P} + \mathrm{PA} & \mathrm{PB} \\ \mathrm{B}^T\mathrm{P} & -\mathrm{I} \end{bmatrix} < 0 \\ \gamma\mathrm{P} - \mathrm{Q} > 0 \end{cases} \tag{12.37}$$

Problem (12.37) is a generalized eigenvalue problem under LMI constraints. It is solved by using the **gevp** MATLAB command as shown in the following example.

MATLAB® exercise 12.9

Consider the system with state-space matrices given by:

$$\mathrm{A} = \begin{bmatrix} -1 & 0 & 0 & 0 & 0 \\ 0 & -2 & 0 & 0 & 0 \\ 0 & 0 & -3 & 0 & 0 \\ 0 & 0 & 0 & -4 & 0 \\ 0 & 0 & 0 & 0 & -5 \end{bmatrix}; \mathrm{B} = \mathrm{C}^T = \begin{bmatrix} 1 \\ 1 \\ 1 \\ 1 \\ 1 \end{bmatrix}$$

and calculate the Hankel norm.
Let's first define the matrices:

```
>> A=diag([-1 -2 -3 -4 -5]);
>> B=ones(5,1);
>> C=B';
```

One of the LMI constraints of the problem is not a strict inequality. To deal with such a constraint, a numerical stratagem has to be adopted by defining a small perturbation matrix:

```
>> m=.00001*eye(5);
```

The LMI problem is now set:

```
>> setlmis([]);
>> P=lmivar(1,[5,1]);
>> Q=lmivar(1,[5,1]);
>> lmiterm([-1 1 1 P],1,1);
>> lmiterm([1 1 1 0],0);
>> lmiterm([-2 1 1 Q],1,1);
>> lmiterm([2 1 1 0],0);
>> lmiterm([3 1 1 P],1,A,'s');
>> lmiterm([3 2 1 P],B',1);
>> lmiterm([3 2 2 0],-1);
>> lmiterm([4 1 1 Q],1,A,'s');
>> h=C'*C+m;
>> lmiterm([-4 1 1 0],-h);
>> lmiterm([5 1 1 Q],1,1);
>> lmiterm([-5 1 1 P],1,1);
>> lmisys=getlmis;
>> options=[0.00001 0 0 0 0];
>> [Hnorm,PQ]=gevp(lmisys,1,options);
>> P=dec2mat(lmisys,PQ,P);
>> Q=dec2mat(lmisys,PQ,Q);
```

One obtains: $\|G(s)\|_H = 1.1151$.
This result can be compared with the calculation of the maximum singular value of the system as follows:

```
>> Wo2=gram(A',C');
>> Wc2=gram(A,B);
>> S=eig(Wo2*Wc2);
>> s1=max(S)
```

which yields $\sigma_1 = 1.1150$.

12.9 H_∞ control

In this paragraph we will show the LMI problem to solve the H_∞ control problem through an example. Consider the system:

$$\begin{aligned} \dot{x} &= w + 2u \\ z &= x \\ y &= -x + w \end{aligned} \tag{12.38}$$

Remember that the general equations for a system at which we want to apply the H_∞ control, are given by:

$$\begin{cases} \dot{\mathbf{x}} = \mathrm{A}\mathbf{x} + \mathrm{B}_1\mathbf{w} + \mathrm{B}_2\mathbf{u} \\ \mathbf{z} = \mathrm{C}_1\mathbf{x} + \mathrm{D}_{11}\mathbf{w} + \mathrm{D}_{12}\mathbf{u} \\ \mathbf{y} = \mathrm{C}_2\mathbf{x} + \mathrm{D}_{21}\mathbf{w} + \mathrm{D}_{22}\mathbf{u} \end{cases} \tag{12.39}$$

The system (12.38) is equal to the system (12.39) if we choose the following values of the parameters: $a = 0$; $b_1 = 1$; $b_2 = 2$; $c_1 = 1$; $d_{11} = d_{12} = 0$; $c_2 = -1$; $d_{21} = 1$; $d_{22} = 0$.

Now let us define in MATLAB the system as a ltisys object

```
>> A=0;
>> B1=1;
>> B2=2;
>> C1=1;
>> D11=0;
>> D12=0;
>> C2=-1;
>> D21=1;
>> D22=0;
>> P=ltisys(A,[B1 B2], [C1; C2],[D11 D12; D21 D22])
```

To solve the problem we can use two methods: the resolution method based on the Riccati equations or the method based on an optimization problem with LMI constraints. In the first case we need to follow the procedure described in Chapter 11. In the second case we have to consider the LMI minimization problem with linear objective defined by the constraints:

$$\begin{pmatrix} N_{12} & 0 \\ 0 & I \end{pmatrix}^T \begin{pmatrix} AR + RA^T & RC_1^T & B_1 \\ C_1R & -\gamma I & D_{11} \\ B_1^T & D_{11}^T & -\gamma I \end{pmatrix} \begin{pmatrix} N_{12} & 0 \\ 0 & I \end{pmatrix} < 0 \qquad (12.40)$$

$$\begin{pmatrix} N_{21} & 0 \\ 0 & I \end{pmatrix}^T \begin{pmatrix} A^TS + SA & SB_1^T & C_1^T \\ B_1^TS & -\gamma I & D_{11}^T \\ C_1 & D_{11} & -\gamma I \end{pmatrix} \begin{pmatrix} N_{21} & 0 \\ 0 & I \end{pmatrix} < 0 \qquad (12.41)$$

$$\begin{pmatrix} R & I \\ I & S \end{pmatrix} \geq 0 \qquad (12.42)$$

where N_{12} and N_{21} are the bases of the null spaces, respectively, of (B_2^T, D_{12}^T) and of (C_2, D_{21}). In MATLAB, we do not need to define the inequalities (12.40)-(12.42) every time, but we can use directly the command `hinflmi`.

To use the method based on the Riccati equations the following command has to be used:

```
>> [gopt,K]=hinfric(P,[1 1])
```

while to use the LMI-based method, the following command has to be used:

```
>> [gopt,K]=hinflmi(P,[1 1])
```

In both cases, the size of matrix matrice D_{22} (i.e., `[1 1]`) or, in other words, the size of the output vector $\mathbf{y}$ and of the input vector $\mathbf{u}$, has to be specified.

Similar performances are obtained with the two approaches ($\gamma = 1.0088$ with the method based on Riccati equations, $\gamma = 1.0023$ with the LMI-based method).

To verify the obtained performances, the control scheme in Figure 11.2 has first to be implemented. This can be done by connecting within a feedback loop system $P(s)$ and controller $K(s)$. If the two systems have been defined as ltisys objects, one has to use the command `slft`; while if the two systems have been defined as lti objects, the command `feedback` has to be used.

Consider the first case. The closed-loop system is defined by:

```
>> systemcl=slft(P,K,1,1)
```

sistemacl also is a ltisys object, i.e., it is defined by the realization matrix. Note that in this case, since u is no more an external input of the system, but it derives from the feedback of y, the closed-loop system has a 2×2 state matrix and $D^{cl} \in \mathbb{R}^{1\times 1}$: $sistemacl = \begin{bmatrix} A^{cl}_{2\times 2} & B^{cl}_{1\times 1} \\ C^{cl}_{1\times 1} & D^{cl}_{1\times 1} \end{bmatrix}$.

To verify, for instance, that the closed-loop system is stable, one has to find matrix A^{cl} and then to compute its eigenvalues:

```
>> eig(systemcl(1:2,1:2))
```

this can be done also by using the command:

```
>> spol(sistemacl)
```

Obviously, closed-loop eigenvalues with negative real part ($\lambda_1 = -0.0726 \cdot 10^4$ and $\lambda_2 = -9.8886 \cdot 10^4$) are obtained.

Otherwise, lti objects can be used. For example, the closed-loop system can be rewritten as a lti object (recall that, in this case, the input is w, and the output is z):

```
>>systemcllti1=ss(systemcl(1:2,1:2),systemcl(1:2,3),
        systemcl(3,1:2),systemcl(3,3))
```

Analogously, the closed-loop system can be defined after having defined system P and controller $K(s)$ as LTI object, and then use the command `feedback`:

```
>> Plti=ss(A,[B1 B2],[C1; C2],[D11 D12; D21 D22]);
>> Klti=ss(K(1,1),K(1,2),K(2,1),K(2,2));
>> systemcllti2=feedback(Plti,Klti,2,2,+1)
```

Note (for example, looking at the `help` of command `feedback`) that, in this case, the system 'systemcllti' has two inputs and two outputs, since the scheme implemented by the command `feedback` takes into account both output z and output y and both input w and input u.

By the plot of the Bode diagram (of the closed-loop system) it can be verified that $\|T_{zw}(s)\|_\infty \leq 1.0023$. The command to be used is

```
>> bode(systemcllti1)
```

or

```
>> bode(systemcllti2(1,1))
```

12.10 Multiobjective control

As discussed above, the objective of the H_∞ control is to find among the controllers that stabilize the closed-loop system the one minimizing the norm $\|T_{zw}(s)\|_\infty$. However, generally, there exist other types of performances that can be optimized. The multiobjective control has been introduced for this reason, allowing to take into account other types of performances, beyond that represented by the H_∞ norm.

The reference scheme of the multiobjective control is shown in Figure 12.2 where further outputs (variables $\mathbf{z}_2$) are highlighted. The state-space equations corresponding to the control scheme of Figure 12.4 are given by:

$$\left\{ \begin{array}{l} \dot{\mathbf{x}} = \mathrm{A}\mathbf{x} + \mathrm{B}_1\mathbf{w} + \mathrm{B}_2\mathbf{u} \\ \mathbf{z}_\infty = \mathrm{C}_1\mathbf{x} + \mathrm{D}_{11}\mathbf{w} + \mathrm{D}_{12}\mathbf{u} \\ \mathbf{z}_2 = \mathrm{C}_2\mathbf{x} + \mathrm{D}_{21}\mathbf{w} + \mathrm{D}_{22}\mathbf{u} \\ \mathbf{y} = \mathrm{C}_3\mathbf{x} + \mathrm{D}_{31}\mathbf{w} + \mathrm{D}_{32}\mathbf{u} \end{array} \right. \tag{12.43}$$

In the multiobjective control what is minimized is the weighted sum of two terms: the first is the H_∞ norm of the transfer matrix from $\mathbf{w}$ to $\mathbf{z}_\infty$; the second is the H_2 norm (defined below) of the transfer matrix from $\mathbf{w}$ to $\mathbf{z}_2$. The idea underlying the multiobjective control will be further discussed, after defining the H_2 norm of a system.

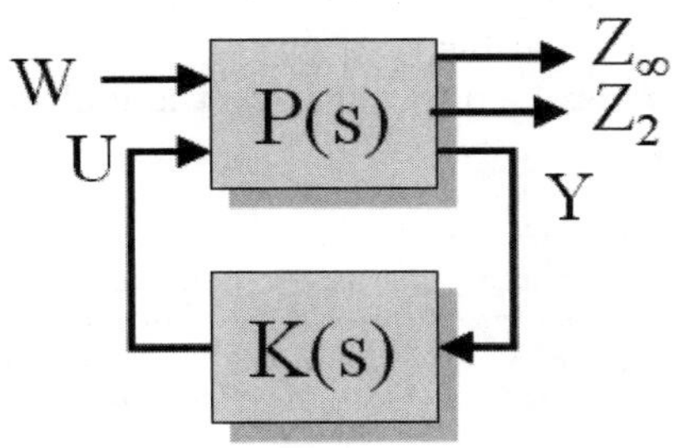

FIGURE 12.2
Scheme of the multiobjective control.

The H_2 norm is defined as follows:

$$\begin{aligned} \|G(s)\|_2 &= \left(\frac{1}{2\pi}\int_{-\infty}^{\infty} \mathtt{trace}[G(j\omega)G(j\omega)^*]d\omega\right)^{\frac{1}{2}} = \\ &= \left(\frac{1}{2\pi}\int_{-\infty}^{\infty}\sum_{i=1}^{r}\sigma_i[G(j\omega)]d\omega\right)^{\frac{1}{2}} \end{aligned} \tag{12.44}$$

where r is the rank of $G(j\omega)$.

For systems with stochastic inputs, the physical meaning of the H_2 norm derives from the fact that, when the system is driven by a white noise with zero mean and unitary variance, then the H_2 norm is equal to:

$$\|G(s)\|_2 = (E[\mathbf{y}^T(t)\mathbf{y}(t)])^{\frac{1}{2}} \tag{12.45}$$

The H_2 norm thus represents a measure of the energy associated to the output of a system driven by a white noise with zero mean and unitary variance.

The H_2 norm can be also viewed as the energy associated to the signal z_2. In fact, for a system given that $\mathbf{y}(t)$ is the impulse response it holds:

$$\|G(s)\|_2 = (\mathtt{trace}[\int_0^\infty \mathbf{y}^T(t)\mathbf{y}(t)])^{\frac{1}{2}} \tag{12.46}$$

As concerns the calculation of the H_2 norm, it can be performed from the controllability and observability gramians:

$$\|G(s)\|_2 = (\mathtt{trace}(\mathrm{C}\mathrm{W}_c^2\mathrm{C}^T))^{\frac{1}{2}} = (\mathtt{trace}(\mathrm{B}^T\mathrm{W}_o^2\mathrm{B}))^{\frac{1}{2}} \tag{12.47}$$

The aim of the multiobjective control is thus to find a controller that stabilizes the closed-loop system and that, indicated with $\mathrm{T}_{z_\infty,w}$ and $\mathrm{T}_{z_2,w}$ the transfer matrices from $\mathbf{w}$ to $\mathbf{z}_\infty$ and from $\mathbf{w}$ to $\mathbf{z}_2$, minimizes $\alpha\|\mathrm{T}_{z_\infty,w}\|_\infty^2 + \beta\|\mathrm{T}_{z_2,w}\|_2^2$ with $\alpha \geq 0$ e $\beta \geq 0$.

With MATLAB the multiobjective control can be run with the command `hinfmix`. This also allows us to include a further constraint (also formulated in terms of LMIs) related to the region in which the closed-loop eigenvalues have to lie. Obviously, not all the multiobjective control are feasible, since the constraints cannot be compatible.

The following examples help to illustrate the multiobjective control.

MATLAB® exercise 12.10

We start again with the example discussed in paragraph 12.9, referring to system (12.38), and add a further specification to the control system: we now also want to minimize the energy associated with the input. To do this, a multiobjective control has to be set by defining a new variable $z_2 = u$. The equations of the system to be controlled become:

$$\begin{aligned} \dot{x} &= w + 2u \\ z_\infty &= x \\ z_2 &= u \\ y &= -x + w \end{aligned} \tag{12.48}$$

System (12.48) is equivalent to system (12.43) with $a = 0$; $b_1 = 1$; $b_2 = 2$; $c_1 = 1$; $d_{11} = d_{12} = 0$; $c_2 = 0$; $d_{21} = 0$; $d_{22} = 1$, $c_3 = -1$; $d_{31} = 1$; $d_{32} = 0$.
Assuming that system P has been defined in MATLAB as a `ltisys` object (as in the examples dealt with previously), we now discuss the use of command `hinfmix`. The syntax is:

```
[gopt,h2opt,K,R,S] = hinfmix(P,rv,obj,region)
```

where P is the process to be controlled (`ltisys` object), rv is a vector listing the sizes of z_2, y and u and $obj = \left[\ VMNormH_\infty \quad VMNormH_2 \quad \alpha \quad \beta\ \right]$ where $VMNormH_\infty$ and $VMNormH_2$ are the maximum admissible values for $\|\mathrm{T}_{z_\infty,w}\|_\infty$ and $\|\mathrm{T}_{z_2,w}\|_2$, respectively, or in other words the minimum performances that have to be guaranteed.
Given this syntax, note that command `hinfmix` can be used in a flexible way. For instance, the H_∞ control can be implemented as follows:

```
>> [gopt,h2opt,K_Hinf]=hinfmix(P,[1 1 1],[0 0 1 0])
```

The performance so obtained is: $\|\mathrm{T}_{z_\infty,w}\|_\infty \leq gopt = 1.0009$. In this case, the control does not take into account in any way the objective defined by the H_2 norm.
To evaluate the performance of the control system, the closed-loop system has first to be calculated:

```
>>systemLTIol=ss(P(1,1),P(1,2:3),P(2:4,1),P(2:4,2:3));
>>compensatorLTI=ss(K_Hinf(1,1),K_Hinf(1,2),
          K_Hinf(2,1),K_Hinf(2,2));
>>systemLTIcl=feedback(systemLTIol,compensatorLTI,2,3,1);
```

The inputs of this system are w and u. The outputs are z_∞, z_2 and $\tilde{y}$. So, to study the performance of the H_∞ control, the transfer function from input 1 (w) to output 1 (z_∞) has to be investigated.
We can plot the Bode diagram and verify that the norm H_∞ is less than gopt:

```
>> Hs=tf(systemLTIcl);
>> bode(Hs(1,1))
>> normhinf(Hs(1,1))
```

Consider now a H_2/H_∞ problem in which the two norms are weighted in the same way ($\alpha = 1$, $\beta = 1$) and in which no minimum specification is considered:

```
>> [gopt,h2opt,K_H2_Hinf]=hinfmix(P,[1 1 1],[0 0 1 1]);
```

In this case, we obtain $gopt = 1.9104$ and $h2opt = 1.2110$. The performance in terms of H_∞ norm is slightly worse, but now the H_2 norm is quite low. This can be verified by plotting the impulse response of the closed-loop system controlled with the H_∞ technique and the one of the system controlled with the H_2/H_∞ technique.
We first define the new closed-loop system by the commands:

```
>> systemLTIol=ss(P(1,1),P(1,2:3),P(2:4,1),P(2:4,2:3));
>>compensatorLTI=ss(K_H2_Hinf(1,1),K_H2_Hinf(1,2),
          K_H2_Hinf(2,1),K_H2_Hinf(2,2));
```

```
>>systemLTIcl2=feedback(systemLTIol,compensatorLTI,2,3,1);
>>Hs2=tf(systemLTIcl2);
```
and then compare the two plots of the impulse response:
```
>> figure(1); impulse(systemLTIcl)
>> figure(2); impulse(systemLTIcl2)
```
In the latter case (H_2/H_∞ control) the variables have smaller amplitudes (and so energy).

The comparison of the Bode diagrams reveals that the H_∞ control leads to a larger bandwidth. Therefore, the system controlled with the H_∞ technique is much faster than that controlled with the H_2/H_∞ technique, but requires more energy to be controlled. Note also that the objectives achieved in this way represent the maximum performances that can be obtained with the H_2/H_∞ control. In fact, if we consider for instance:
```
>> [gopt,h2opt,K_H2_Hinf]=hinfmix(P,[1 1 1],[1 1 1 1]);
```
we find that the problem is not feasible.

Finally, we show an example where a specification on the region in which the closed-loop eigenvalues have to lie is also included. In fact note that the closed-loop eigenvalues of the H_∞ control, obtained by
```
>> pole(Hs)
```
are quite large: $\lambda_1 = -1.4087 \cdot 10^4$, $\lambda_2 = -0.2357 \cdot 10^4$, $\lambda_3 = -1.4087 \cdot 10^4$, $\lambda_2 = -0.2357 \cdot 10^4$.

Let us therefore include a new constraint. We want the closed-loop eigenvalues to be such that $-10 \leq Re\lambda \leq 0$. By using the command
```
>> region=lmireg
```
through a series of multiple choices the given region can be set. Then, by using the command:
```
>>[gopt,h2opt,K_H2_Hinf_R]=hinfmix(P,[1 1 1],[0 0 1 0],region);
```
the H_∞ control with a constraint on the position of the closed-loop eigenvalues is defined.

We now calculate the transfer matrix obtained with this control and indicate it as $Hs3$:
```
>>systemLTIol=ss(P(1,1),P(1,2:3),P(2:4,1),P(2:4,2:3));
>>compensatorLTI=ss(K_H2_Hinf_R(1,1),K_H2_Hinf_R(1,2),
     K_H2_Hinf_R(2,1),K_H2_Hinf_R(2,2));
>>systemLTIcl3=feedback(systemLTIol,compensatorLTI,2,3,1);
>>Hs3=tf(systemLTIcl3);
```
note that now the closed-loop eigenvalues
```
>> pole(Hs3)
```
are much smaller $\lambda_{1,2} = \lambda_{3,4} = -8.9068 + j8.3754$.

By comparing the Bode diagrams and the impulse responses of the controlled systems obtained with the three different techniques we note that in this latter case we have intermediate performances. The Bode diagrams of the given example are shown in Figure 12.3.

MATLAB® exercise 12.11

As a second example of multiobjective control, consider the system shown in Figure 12.4. The process to be controlled is a first order system with transfer function $G(s) = \frac{1}{s+1}$. The first objective of the control system (beyond closed-loop stability) is to track the input, i.e., to minimize the error $e(t) = r(t) - y(t)$.

First, formulate the control problem in terms of the block scheme of the multiobjective control. To do this, let's write the state-space equations by introducing the variable $\tilde{y} = r - x$ (that is the input of the controller):

$$\begin{aligned} \dot{x} &= -x + u \\ \tilde{y} &= r - x \end{aligned} \tag{12.49}$$

Define the variables $\mathbf{z}$: to satisfy the objective of input tracking, fix $z_\infty = e$ and $w = r$. In this way, in fact, we consider the worst case for the error. In fact, for a SISO system

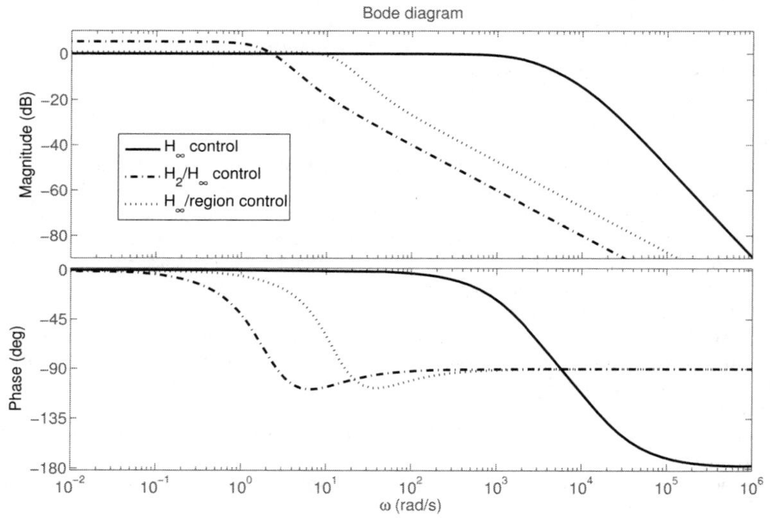

FIGURE 12.3
Performances obtained by the H_∞ control technique, the H_2/H_∞ control technique and the H_∞ control technique with constraint on the position of the closed-loop eigenvalues. The control is applied to system (12.48).

the H_∞ norm represents the maximum value of the frequency response, or in other words the maximum amplification for any sinusoidal input.
Let's now consider the objective of the H_2 control. We define $z_2 = y = x$ in such a way to minimize the output energy.
The whole equations of the system to be controlled are thus given by:

$$\begin{aligned}\dot{x} &= -x + u \\ z_\infty &= w - x \\ z_2 &= x \\ \tilde{y} &= w - x\end{aligned} \tag{12.50}$$

Comparing equations (12.50) with the reference model for multiobjective control as in equations (12.43), we get: $A = -1$; $B_1 = 0$; $B_2 = 1$; $C_1 = -1$; $D_{11} = 1$; $D_{12} = 0$; $C_2 = 1$; $D_{21} = 0$; $D_{22} = 0$; $C_3 = -1$; $D_{31} = 1$; $D_{32} = 0$.
Let's assume that the control objective is to minimize $5\|\mathrm{T}_{z_\infty,w}\|_\infty^2 + 5\|\mathrm{T}_{z_2,w}\|_2^2$ with the further constraint that the closed-loop eigenvalues are such that $-10 \leq Re\lambda < 0$.
In this case we have to use the command:

```
>> [gopt,h2opt,K]=hinfmix(P,[1 1 1],[1 1 5 5],region)
```

The performances obtained in this case are `gopt` $= 1.0000$ and `h2opt` $= 0.6494$, while the controller is given by $K(s) = \left[\begin{array}{c|c} -5.3574 & 1.5066 \\ \hline -0.6449 & 0.3396 \end{array}\right]$.

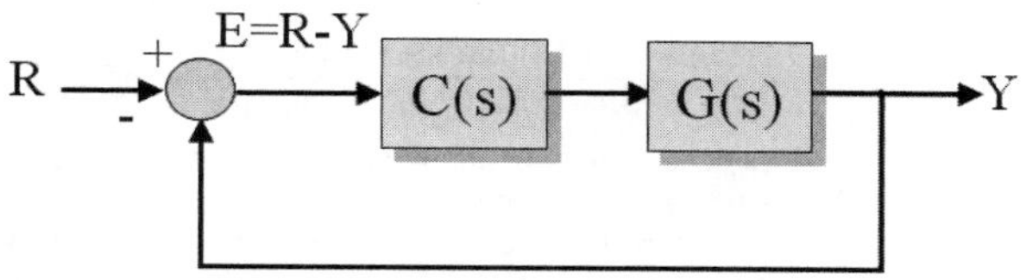

FIGURE 12.4
Example of multiobjective control.

MATLAB® exercise 12.12

In this exercise a further example of H_∞ control is given. Let us consider the problem of controlling the vertical position of plasma in the Joint European Torus (JET) (see for more information the paper by Fortuna et *al.* listed in the references of the chapter). The plasma vertical position can be modeled as a cascade of two stages: a first linear system modeling the Poloidal Radial Field Amplifier (PRFA) with transfer function

$$G_1(s) = \frac{100}{1 + sT_a} \tag{12.51}$$

and the plant model

$$G_2(s) = \frac{1}{sT_v} \frac{1}{s - \gamma} \tag{12.52}$$

where $T_v = 10.4ms$ and $\gamma = 80$.
Consider now the following state-space realization:

$$\mathrm{A} = \begin{bmatrix} 0 & 1 \\ 0 & 80 \end{bmatrix}; \mathrm{B} = \begin{bmatrix} 0 \\ 1 \end{bmatrix}; \mathrm{C} = \begin{bmatrix} 9615 & 0 \end{bmatrix}; \mathrm{D} = 0;$$

Consider again the control scheme reported in Figure 12.4 for which the first objective of the control is to track the input, i.e., to minimize the error $e(t) = r(t) - y(t)$.
Rewrite the state space equations by introducing the variable $\tilde{y} = r - y$:

$$\begin{aligned} \dot{\mathrm{X}} &= \mathrm{AX} + \mathrm{B}u \\ \tilde{y} &= w - \mathrm{CX} \end{aligned} \tag{12.53}$$

To satisfy the objective of input tracking, we choose $z_\infty = e$ and $w = r$. The whole equations of the system to be controlled are rewritten as:

$$\begin{aligned} \dot{\mathrm{X}} &= \mathrm{AX} + \mathrm{B}u \\ z_\infty &= w - \mathrm{CX} \\ \tilde{y} &= w - \mathrm{CX} \end{aligned} \tag{12.54}$$

from which it is possible to define $\mathrm{B}_1 = \begin{bmatrix} 0 & 0 \end{bmatrix}^T$, $\mathrm{B}_2 = \begin{bmatrix} 0 & 1 \end{bmatrix}^T$, $\mathrm{C}_1 = \mathrm{C}_2 = -\mathrm{C}$, $D_{11} = D_{21} = 1$, and $D_{12} = D_{22} = 0$.
The realization matrix of the process can be written as:

$$P(s) = \left[\begin{array}{cc|cc} 0 & 1 & 0 & 0 \\ 0 & 80 & 0 & 1 \\ \hline -9615 & 0 & 1 & 0 \\ -9615 & 0 & 1 & 0 \end{array} \right];$$

Applying the procedure to obtain the H_∞ controller we obtain a performance index $\gamma_{opt} = 1.31$ in correspondence of the following transfer function:

$$K(s) = \frac{1.54 \cdot 10^7 s + 153.2}{(s^2 + 4.70 \cdot 10^4 s + 9.28 \cdot 10^8)}$$

However, using controller $K(s)$ the poles of the closed-loop system are $p_1 = -1.8 \cdot 10^{-5}$, $p_2 = -80$ and $p_{3,4} = -2.34 \cdot 10^4 \pm 1.93 \cdot 10^4 j$.
If we want to design the H_∞ controller choosing a suitable region for the closed-loop system poles, we may use the procedure defined for the multiobjective control imposing that the closed-loop poles have both real and imaginary part between 0 and -20, i.e., the feasible region is a square. In this case the H_∞ controller with a $\gamma_{opt} = 110$ is defined by:

$$K(s) = 0.99\frac{(s + 0.16)}{(s + 114.2)}$$

leading to the following closed-loop poles: $p_1 = -15.52$, $p_2 = -13.57$ and $p_{3,4} = -9.03 \pm 3.62j$.

The LMI techniques discussed can stimulate the students to formulate other control problems, using this powerful and efficient tool.

The Lagrange's equation provides a method to derive the differential equations governing a system. Moreover, the differential Riccati equation is a design tool to obtain the optimal control law, but other matrix differential equations also arise, as shown in the book of Helmke and Moore, in solving and conceiving more system analysis and design tools. The solution of such problems require efficient numerical iterative solving procedures.

Since the formulation of more system theory and modern control optimal problems in terms of LMIs leads to convex optimization problems, it allows to use the interior point algorithm which is a universal solver for many problems and requires few iterations to converge to the solution. Therefore, from modelling to design, differential equations are the core of system analysis and control and efficient iterative optimization procedures are the machinery to solve them.

12.11 Exercises

1. Using LMI techniques, determine the H_∞ norm of the system with transfer function $G(s) = \frac{s+1}{(s+2)(s+3)}$.
2. Using LMI techniques, determine if the system with transfer function $G(s) = 10\frac{s+1}{(s^2+5s+10)}$ is bounded real or not.
3. Given the system:

$$\begin{cases} \dot{\mathbf{x}} = \mathrm{A}\mathbf{x} + \mathrm{B}_1\mathbf{w} + \mathrm{B}_2\mathbf{u} \\ \mathbf{z}_\infty = \mathrm{C}_1\mathbf{x} + \mathrm{D}_{11}\mathbf{w} + \mathrm{D}_{12}\mathbf{u} \\ \mathbf{z}_2 = \mathrm{C}_2\mathbf{x} + \mathrm{D}_{21}\mathbf{w} + \mathrm{D}_{22}\mathbf{u} \\ \mathbf{y} = \mathrm{C}_3\mathbf{x} + \mathrm{D}_{31}\mathbf{w} + \mathrm{D}_{32}\mathbf{u} \end{cases}$$

with $A = \begin{bmatrix} 0 & 1 \\ 0.2 & -1 \end{bmatrix}$; $B_1 = \begin{bmatrix} 0 \\ 1 \end{bmatrix}$; $B_2 = \begin{bmatrix} 0 \\ 1 \end{bmatrix}$; $C_1 = \begin{bmatrix} 2 & 1 \end{bmatrix}$; $C_2 = \begin{bmatrix} 0 & 0 \end{bmatrix}$; $C_3 = \begin{bmatrix} 1 & 0 \end{bmatrix}$; $D_{11} = D_{12} = 0$; $D_{21} = 0$; $D_{22} = 1$; $D_{31} = 1$ and $D_{32} = 0$, design a multiobjective control and verify the obtained performance. Consider then the objective $\|T_{z_\infty w}\|_\infty \leq 0.5$ and $\|T_{z_2 w}\|_2 \leq 2.5$. Is it possible to design a controller satisfying the specifications?

4. Calculate the compensator (regulator and observer) which stabilizes the system

$$\begin{aligned} \dot{\mathbf{x}} &= A\mathbf{x} + B\mathbf{u} \\ \mathbf{y} &= C\mathbf{x} \end{aligned}$$

with $A = \begin{bmatrix} \lambda_1 & 0 & 0 \\ 0 & \lambda_2 & 0 \\ 0 & 0 & \lambda_3 \end{bmatrix}$; $B = \begin{bmatrix} b_{1,i} \\ b_2 \\ b_3 \end{bmatrix}$; $C = B^T$; $\lambda_1 = -1$; $\lambda_2 = 1$; $\lambda_3 = 5$; $b_{1,1} = 4$; $b_{1,2} = -1$; $b_2 = 1$ and $b_3 = 1$.

5. Given the system

$$\begin{cases} \dot{\mathbf{x}} = A\mathbf{x} + B_1\mathbf{w} + B_2\mathbf{u} \\ \mathbf{z}_\infty = C_1\mathbf{x} + D_{11}\mathbf{w} + D_{12}\mathbf{u} \\ \mathbf{z}_2 = C_2\mathbf{x} + D_{21}\mathbf{w} + D_{22}\mathbf{u} \\ \mathbf{y} = C_3\mathbf{x} + D_{31}\mathbf{w} + D_{32}\mathbf{u} \end{cases}$$

with $A = \begin{bmatrix} 0 & 1 \\ 0.2 & -1 \end{bmatrix}$; $B_1 = \begin{bmatrix} 0 \\ 1 \end{bmatrix}$; $B_2 = \begin{bmatrix} 0 \\ 1 \end{bmatrix}$; $C_1 = \begin{bmatrix} 2 & 1 \end{bmatrix}$; $C_2 = \begin{bmatrix} 0 & 0 \end{bmatrix}$; $C_3 = \begin{bmatrix} 1 & 0 \end{bmatrix}$; $D_{11} = D_{12} = 0$; $D_{21} = 0$; $D_{22} = 1$; $D_{31} = 1$ and $D_{32} = 0$, design a multiobjective controller (report the values of the optimal performance) and verify the performance obtained. Then consider the objective $\|T_{z_\infty w}\|_\infty \leq 0.5$ and $\|T_{z_2 w}\|_2 \leq 2.5$. Is it possible to design a controller with these specifications?

6. Using the LMI approach, find the control law that stabilizes simultaneously the two systems:

$A_1 = \begin{bmatrix} -2 & 0 & 0 \\ 0 & 1 & 0 \\ 0 & 0 & 5 \end{bmatrix}$; $B_1 = \begin{bmatrix} 0 \\ 1 \\ 1 \end{bmatrix}$ e $A_2 = \begin{bmatrix} -5 & 0 & 0 \\ 0 & -6 & 0 \\ 0 & 0 & 3 \end{bmatrix}$; $B_2 = \begin{bmatrix} 0 \\ 1 \\ 3 \end{bmatrix}$.

13

The class of stabilizing controllers

CONTENTS

In this chapter we will deal with the problem of determining, rather than a single controller that meets the assigned specifications, a class of controllers that meet the specifications. Using this technique the class of the controllers can be expressed in function of one or more parameters that allow the designer to specify later additional control objectives. For this reason the problem to find a universal class of controllers which satisfy certain specifications takes the name of parameterization of the controllers. For simplicity, we mainly consider SISO systems, even if the discussed results may be easily generalized to MIMO systems as briefly shown in Section 13.5.

13.1 Parameterization of stabilizing controllers for stable processes

Referring to Figure 13.1, which shows the classic feedback control scheme, we want to determine the whole class of controllers which internally stabilize the closed-loop system. With internally stable, we mean that at any point of the feedback loop the system must be stable. The objective is that the system defined by the state variables x_1, x_2 and x_3 is stable, i.e., that all the possible transfer functions from any input of the system (r, d and n) to any of the variables x_1, x_2 and x_3 are stable.

In this paragraph we consider the case in which the process to control is

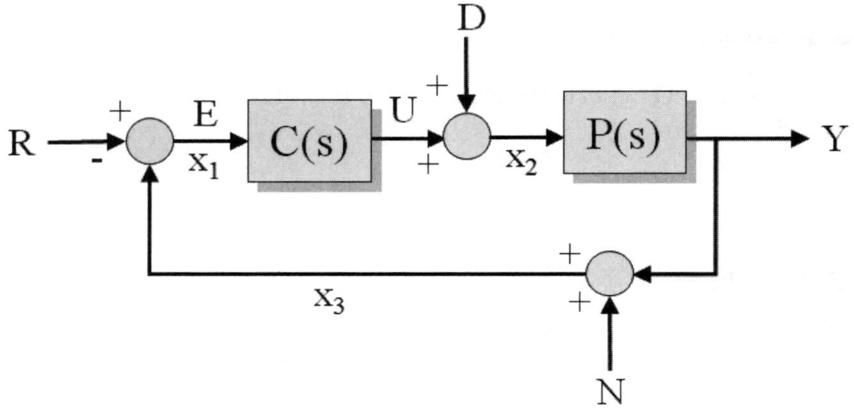

FIGURE 13.1
Feedback control scheme.

asymptotically stable. In the next paragraph we will consider the extension to the case in which the process is not stable. Let $\mathbb{H}^\infty$ indicate the set of asymptotically stable processes. The following theorem expresses the class of stabilizing controllers.

Theorem 25 *Given a process $P(s) \in \mathbb{H}^\infty$ the class of the controllers that make the closed-loop system internally stable is given by:*

$$C(s) = \frac{Q(s)}{1 - P(s)Q(s)} \tag{13.1}$$

with $Q(s) \in \mathbb{H}^\infty$.

Applying the parameterization of stabilizing controllers expressed by the equation (13.1) we obtain a transfer function given by:

$$F(s) = \frac{C(s)P(s)}{1 + C(s)P(s)} = \frac{\frac{Q(s)}{1-P(s)Q(s)}P(s)}{1 + \frac{Q(s)}{1-P(s)Q(s)}P(s)} = P(s)Q(s)$$

From the expression of the closed-loop transfer function we deduce immediately that, if $P(s)$ and $Q(s)$ are asymptotically stable, also $F(s)$ is. Since the controller (13.1) is an internally stabilizing controller, all nine possible transfer functions between r, d and n and x_1, x_2 and x_3 are asymptotically stable.

We can also get immediately that $Q(s)$ is the transfer function between r and u: $Q(s) = \frac{U(s)}{R(s)}$.

Once the parameterization of stabilizing controllers has been found, it is possible to search among the parameterized controllers for the controller (or the controllers) that satisfy further specifications. Consider, for example, the zero error specification of the step response. To impose this specification, we must ensure that the response at the closed-loop unit step is equal to one. Applying the final value theorem (that it is possible to apply, because the closed-loop system is asymptotically stable), we obtain:

$$\lim_{t\to\infty} y(t) = \lim_{s\to 0} sQ(s)P(s)\frac{1}{s} = Q(0)P(0) \tag{13.2}$$

So the condition to impose is

$$Q(0)P(0) = 1$$

Example 13.1

Determine a controller internally stabilizing and that guarantees zero error of the response to the ramp input for the system $P(s) = \frac{1}{(s+1)(s+5)}$.

Consider $Q(s) = \frac{as+b}{s+1}$. Since $Q(0) = b$ and $P(0) = \frac{1}{5}$, imposing that the step error is null (i.e., that $Q(0) = \frac{1}{P(0)}$), we obtain $b = 5$.

To impose that also the ramp error (indicated with $E_R(t)$) is null, consider

$$\lim_{t\to\infty} E_R(t) = \lim_{s\to 0} s\left(\frac{1}{s^2} - \frac{1}{s^2}F(s)\right) = \lim_{s\to 0}\frac{1}{s}(1 - F(s)) = 0$$

$$\Rightarrow \lim_{s\to 0}\frac{1}{s}\left(1 - \frac{as+5}{s+1}\frac{1}{(s+1)(s+5)}\right) = 0$$

$$\Rightarrow \lim_{s\to 0}\frac{s^2 + 7s + (11-a)}{(s+1)^2(s+5)} = \frac{11-a}{5} = 0 \Rightarrow a = 11$$

So $Q(s) = \frac{11s+5}{s+1}$, from which we get that $C(s) = \frac{Q(s)}{1-Q(s)P(s)} = \frac{11s^3+71s^2+85s+25}{s^2(s+7)}$. As we expected, $C(s)$ has two poles at the origin, which are necessary to have zero ramp error.

13.2 Parameterization of stabilizing controllers for unstable processes

Now consider the case of an unstable process. In this case the function $P(s)$ is rewritten as the ratio of two transfer functions that have peculiar properties. In particular, we use a factorization that takes the name of coprime factorization.

Definition 19 *Given an unstable process $P(s)$ we say that we have a coprime factorization $P(s) = \frac{N(s)}{M(s)}$, if $N(s)$ and $M(s)$ are stable functions and there exist two transfer functions $X(s)$ and $Y(s)$ so that $\forall s$ the following relation holds*

$$N(s)X(s) + M(s)Y(s) = 1 \tag{13.3}$$

Before enunciating which is the necessary and sufficient condition for a factorization to be coprime, we show with an example a factorization that cannot satisfy the equation (13.3).

Example 13.2

Consider the process $P(s) = \frac{1}{s-1}$ and the factorization $N(s) = \frac{1}{(s+1)^2}$ and $M(s) = \frac{s-1}{(s+1)^2}$. The two functions $N(s)$ and $M(s)$ are asymptotically stable and $P(s) = \frac{N(s)}{M(s)}$. The condition (13.3) becomes

$$\frac{1}{(s+1)^2}X(s) + \frac{s-1}{(s+1)^2}Y(s) = 1$$

and has to hold $\forall s$. In particular, considering the limit case $s \to \infty$, the condition to satisfy becomes $0 \cdot X + 0 \cdot Y = 1$, which is clearly impossible to solve. So the factorization is not coprime.

In this last example we can note that $N(s)$ has two zeros at infinity, while $M(s)$ has one zero at infinity and one zero at the unstable pole of $P(s)$. In fact, the examined factorization is not coprime, just because its functions have a common zero (at infinity). Generalizing this result, we can conclude that, whenever there exists a s_0 such that $N(s_0) = M(s_0) = 0$ (i.e., whenever $N(s)$ and $M(s)$ have a common zero), the factorization is not coprime. In fact, this condition is a necessary and sufficient condition, as summarized in the following theorem.

Theorem 26 *Necessary and sufficient condition for a factorization to be coprime is that $N(s)$ and $M(s)$ have no common zero neither finite nor infinite.*

It follows that a coprime factorization of $P(s) = \frac{1}{s-1}$ is given by: $N(s) = \frac{1}{(s+1)}$ and $M(s) = \frac{s-1}{(s+1)}$.

The coprime factorization permits us to express the class of the stabilizing controllers for unstable processes.

Theorem 27 *Given a process $P(s)$, generally unstable, the class of the controllers that make the closed-loop system internally stable is given by:*

$$C(s) = \frac{X(s) + M(s)Q(s)}{Y(s) - N(s)Q(s)} \tag{13.4}$$

where $Q(s) \in \mathbb{H}^\infty$ and $P(s) = \frac{N(s)}{M(s)}$ with $N(s) \in \mathbb{H}^\infty$ and $M(s) \in \mathbb{H}^\infty$ being a coprime factorization of $P(s)$.

With the parameterization (13.4) we obtain a closed-loop function equal to

$$F(s) = \frac{C(s)P(s)}{1 + C(s)P(s)} = \frac{(X(s) + M(s)Q(s))N(s)}{M(s)Y(s) + X(s)N(s)} = (X(s) + M(s)Q(s))N(s)$$

Also in this case we can immediately say that, if $Q(s) \in \mathbb{H}^\infty$, then also $F(s) \in \mathbb{H}^\infty$.

The coprime factorization can also be done in time domain. Consider a system in minimum form $\mathrm{R} = \begin{bmatrix} \mathrm{A} & \mathrm{B} \\ \mathrm{C} & \mathrm{D} \end{bmatrix}$.

The functions $M(s)$ and $N(s)$ are given by

$$M(s) = \left[\begin{array}{c|c} \mathrm{A} + \mathrm{BF} & \mathrm{B} \\ \hline \mathrm{F} & \mathrm{I} \end{array}\right] \tag{13.5}$$

$$N(s) = \left[\begin{array}{c|c} \mathrm{A} + \mathrm{BF} & \mathrm{B} \\ \hline \mathrm{C} + \mathrm{DF} & \mathrm{D} \end{array}\right] \tag{13.6}$$

Note that in both realizations of $N(s)$ and $M(s)$ the state matrix $\mathrm{A}^* = \mathrm{A} + \mathrm{BF}$ is the same. In fact, to factor $P(s)$, the two functions $N(s)$ and $M(s)$ must have the same poles. Moreover, since $N(s) \in \mathbb{H}^\infty$ and $M(s) \in \mathbb{H}^\infty$, we need to choose F so that A^* has eigenvalues with negative real part.

Once matrix F has been found solving a problem of eigenvalue placement, $N(s)$ and $M(s)$ can be obtained through the following expressions:

$$M(s) = \mathrm{F}\left(s\mathrm{I} - \mathrm{A} - \mathrm{BF}\right)^{-1}\mathrm{B} + \mathrm{I}$$

$$N(s) = \left(\mathrm{C} + \mathrm{DF}\right)\left(s\mathrm{I} - \mathrm{A} - \mathrm{BF}\right)^{-1}\mathrm{B} + \mathrm{D}$$

To find $X(s)$ and $Y(s)$ we have to solve another problem of eigenvalue placement. In particular, given that

$$X(s) = \left[\begin{array}{c|c} \mathrm{A} + \mathrm{HC} & \mathrm{H} \\ \hline \mathrm{F} & 0 \end{array}\right] \tag{13.7}$$

$$Y(s) = \left[\begin{array}{c|c} \mathrm{A} + \mathrm{HC} & -\mathrm{B} - \mathrm{HD} \\ \hline \mathrm{F} & \mathrm{I} \end{array}\right] \tag{13.8}$$

we need to find H through eigenvalue placement (chosen arbitrarily). Given the particular structure of $X(s)$ and $Y(s)$ we can prove that they are such that the identity $NX + MY = I$ is satisfied.

Example 13.3

Given the system $P(s) = \frac{1}{(s-2)(s-1)}$, design a compensator, such that to stabilize the closed-loop system, assuring zero error of the step response and assuring that the effect of the disturbance $d(t) = A\sin(\omega t)$ with $\omega = 10rad/s$ is null for $t \to \infty$.

At first, since the process $P(s)$ is unstable, we set a coprime representation. For example we can choose $N(s) = \frac{1}{(s+1)^2}$ and $N(s) = \frac{(s-2)(s-1)}{(s+1)^2}$. Imposing that $N(s)X(s) +$

$M(s)Y(s) = 1$, $X(s) = \frac{19s-11}{s+1}$ and $Y(s) = \frac{s+6}{s+1}$ (functions chosen arbitrary) can be obtained.
To ensure that the error of the step response is null, we must impose that $\lim_{s\to 0} F(s) = 1$ and so that

$$\lim_{s\to 0} N(s)(X(s) + M(s)Q(s)) = 1$$

$$\Rightarrow N(0)X(0) + N(0)M(0)Q(0) = 1$$

$$\Rightarrow Q(0) = 6$$

Finally, consider the specification of the disturbance. From the scheme in Figure 13.1 we can get the transfer function from disturbance to output: $\frac{Y(s)}{D(s)} = \frac{P(s)}{1+C(s)P(s)} = N(s)\,(Y(s) - N(s)Q(s))$. To ensure that the effect of the sinusoidal disturbance is null, we must impose that for $\omega = 10 rad/s$ we have $N(j\omega)\,(Y(j\omega) - N(j\omega)Q(j\omega)) = 0$ $\Rightarrow Q(j\omega) = -94 + 70j$.
Writing $Q(s)$ as $Q(s) = c_1 + \frac{c_2}{s+1} + \frac{c_3}{(s+1)^2}$, we find $Q(s)$ imposing that:

$$Q(0) = c_1 + c_2 + c_3 = 6$$

and

$$Q(j\omega) = c_1 + \frac{c_2}{1+10j} + \frac{c_3}{(1+10j)^2} = 0$$

Equating the real part and the imaginary part of the first and second member of the last relation, we find that $Q(s) = \frac{-79s^2-881s+6}{(s+1)^2}$.

13.3 Parameterization of stable controllers

In this paragraph we will consider a further very important specification. In the parameterizations seen previously there is no guarantee that the compensator $C(s)$ is stable. Consider now under which conditions it is possible to stabilize a process $P(s)$ through stable compensators. Systems in which it is possible to find a stable and stabilizing controller are called strongly stabilizable systems.

Definition 20 *A system that can be stabilized with a stable compensator is called strongly stabilizable.*

Suppose that the process $P(s)$ has zeros with positive real part, i.e., it is a minimum phase system. For example, consider $P_1(s) = \frac{(s-1)(s-5)}{(s+1)^2}$ and look if it is possible to stabilize $G(s)$ with a compensator of type $C(s) = k$. The root locus, i.e., the locus of the points such that $kP_1(s) + 1 = 0$ varying k, shows the positions of the closed-loop poles with respect to k and permits us to understand if there exists a value of k for which $C(s) = k$ is a stable and stabilizable compensator. The root locus of the system $P_1(s) = \frac{(s-1)(s-5)}{(s+1)^2}$ is shown in Figure 13.2. As you can note, for large enough values of k we find stable and stabilizable compensators. Vice versa, consider the system

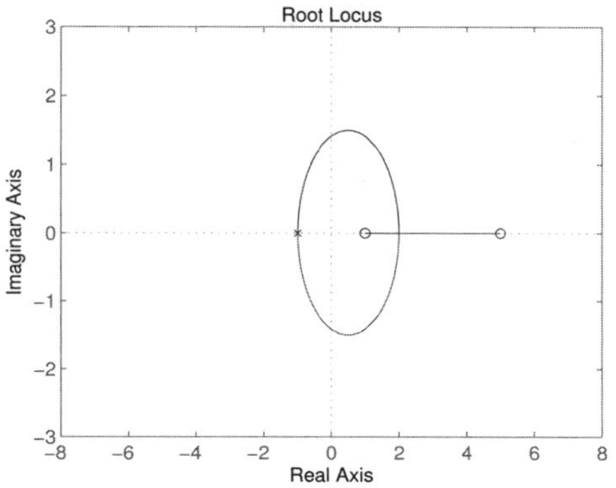

FIGURE 13.2
Root locus of the system $P_1(s) = \frac{(s-1)(s-5)}{(s+1)^2}$.

$P_2(s) = \frac{(s-1)(s-5)}{(s+1)(s-2)}$. Its root locus is shown in Figure 13.3. In this case a controller of type $C(s) = k$ which stabilizes the system does not exist.

The difference between the two systems $P_1(s)$ and $P_2(s)$ is that in the second case between two zeros with positive real part there is an odd number of poles. This condition that we have illustrated for $C(s) = k$ is actually a more general property which guarantees the existence of stable and stabilizing controllers, as expressed in the following theorem.

Theorem 28 *The necessary and sufficient condition so that a system $P(s)$ is strongly stabilizable is that between any pair of zeros with positive real part there is an even number of poles.*

This condition is called Parity Interlacing Property and it has to be verified taking into account also the zeros at infinity.

If a system satisfies the Parity Interlacing Property then it is possible to find stable and stabilizing controllers. Before introducing the parameterization of those controllers we need to define what is meant by unit function.

Definition 21 *A function $Q(s) \in \mathbb{H}^\infty$ is said to be unit if also $\frac{1}{Q(s)} \in \mathbb{H}^\infty$.*

For example, the function $Q(s) = \frac{1}{s+1}$ is not unit, as its inverse $\frac{1}{Q(s)} = s+1$ is not realizable. Instead, $Q(s) = \frac{s+5}{s+2}$ is unit, while $Q(s) = \frac{s-5}{s+2}$ is not unit, because its inverse does not belong to $\mathbb{H}^\infty$.

The parameterization of the class of stable and stabilizing compensators is expressed through the following theorem.

Theorem 29 *Given a process $P(s) = \frac{N(s)}{M(s)}$ with $N(s)$ and $M(s)$ being a*

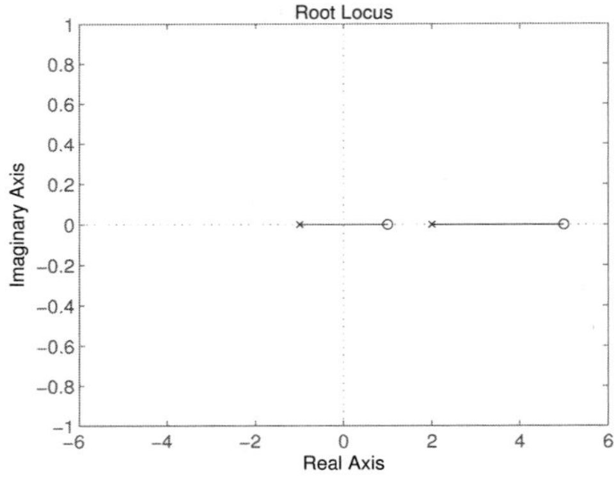

FIGURE 13.3
Root locus of the system $P_2(s) = \frac{(s-1)(s-5)}{(s+1)(s-2)}$.

coprime factorization of $P(s)$, the class of controllers which are stable and stabilizing the closed-loop system is given by:

$$C(s) = \frac{U(s) - M(s)}{N(s)} \tag{13.9}$$

where $U(s)$ is a unit function.

Example 13.4

Consider for example $P(s) = \frac{s-1}{s(s-0.5)}$. It is immediate to see that $P(s)$ verifies the Parity Interlacing Property, as between the zero with positive real part $z = 1$ and the zero at infinity there are no poles. The system $P(s)$ can be stabilized using the compensator (13.9).
Given the coprime factorization $N(s) = \frac{s-1}{(s+1)^2}$ and $M(s) = \frac{s(s-0.5)}{(s+1)^2}$, $C(s)$ is given by:

$$C(s) = \frac{U(s) - \frac{s(s-0.5)}{(s+1)^2}}{\frac{s-1}{(s+1)^2}} = \frac{(s+1)^2U(s) - s(s-0.5)}{s-1} \tag{13.10}$$

In order to obtain a stable $C(s)$, the term $s-1$ at the denominator must be simplified. So, in correspondence of the zero with positive real part ($z = 1$), the numerator of $C(s)$ must be zero. So, we have to impose that $U(1) - M(1) = 0$, i.e., $U(1) = M(1)$.
A second condition to impose derives from the fact that, choosing for simplicity $U(s)$ of first order, the degree of the numerator of $C(s)$ has to be equal to one. In fact the order of $C(s)$ is $n = 1$ and the numerator, to ensure that the system is realizable, must also be a polynomial of degree one. For this, the maximum degree coefficient of the term $(s+1)^2U(s)$ and of the term $s(s-0.5)$ must be equal, so that the numerator of $(s+1)^2U(s) - s(s-0.5)$ is a polynomial of degree two. Remember also that this term has a root $z = 1$ that simplifies then with the term $s-1$: this makes $C(s)$ realizable.
To impose the condition that the maximum degree coefficients of these two terms are equal, we must make sure that $U(\infty) = M(\infty)$ since, for example, $\lim_{s\to\infty} M(s)$ is exactly the value of the maximum degree coefficient.
So we get two interpolation conditions:

$$U(1) = M(1) \tag{13.11}$$

$$U(\infty) = M(\infty) \tag{13.12}$$

Assuming that $U(s) = k\frac{s+\alpha}{s+\beta}$ (unit function if $\alpha > 0$ and $\beta > 0$) and imposing these two conditions we find that

$$k\frac{1+\alpha}{1+\beta} = \frac{1}{8} \tag{13.13}$$

$$k = 1 \tag{13.14}$$

It follows that the parameters of $U(s)$ should be on the line with equation $\frac{1+\alpha}{1+\beta} = \frac{1}{8}$. All the points on this line are suitable values to obtain stable and stabilizing controllers. For example choosing $\alpha = 1$, we have $\beta = 15$. So we have $U(s) = \frac{s+1}{s+15}$ (unit since $U(s) \in \mathbb{H}^\infty$ and $U^{-1}(s) \in \mathbb{H}^\infty$) and $C(s) = -\frac{11.5s+1}{s+15}$.

13.4 Simultaneous stabilizability of two systems

The techniques introduced in this chapter allow us to establish the conditions for the solvability of the problem of simultaneous stability of the two systems. Consider the reference scheme in Figure 13.1 and suppose to want to find a compensator that stabilizes simultaneously two systems, $P_1(s)$ and $P_2(s)$. The conditions of simultaneous stabilizability depend on the system characteristics defined by $\Delta(s) = P_1(s) - P_2(s)$. The main result is given by the following theorem.

Theorem 30 *Given the two systems $P_1(s)$ and $P_2(s)$ and defined $\Delta(s) = P_1(s) - P_2(s)$, it is possible to find a compensator that stabilizes both systems if $\Delta(s)$ satisfies the Parity Interlacing Property.*

As you can note the result is only valid for the simultaneous stabilizability of two systems, whereas the technique shown in Chapter 12 can be applied on two or more systems, but it does not allow to establish if the compensator exists.

Also note that the compensator stabilizing both systems is the stable compensator that stabilizes the system $\Delta(s)$.

13.5 Coprime factorizations for MIMO systems and unitary factorization

Since matrix multiplication is not commutative, for MIMO systems we have to distinguish between left and right coprime factorization. Given the MIMO system $P(s) = N(s)M^{-1}(s)$, the factorization is right coprime if

$$X(s)N(s) + Y(s)M(s) = I$$

In this case the class of the controllers that make the closed-loop system internally stable (which for SISO systems is given by equation (13.4)) is given by:

$$C(s) = (Y(s) - Q(s)N(s))^{-1}(X(s) + Q(s)M(s)) \tag{13.15}$$

Instead, given $P(s) = \tilde{M}^{-1}(s)\tilde{N}(s)$, the factorization is left coprime if

$$\tilde{N}(s)\tilde{X}(s) + \tilde{M}(s)\tilde{X}(s) = I$$

and the class of the controllers that make the closed-loop system internally stable is given by:

$$C(s) = (\tilde{X}(s) + \tilde{M}(s)Q(s))(\tilde{Y}(s) - \tilde{N}(s)Q(s))^{-1} \tag{13.16}$$

In the coprime factorization obtained through the time domain method we have seen that it is possible to choose in an arbitrary way the eigenvalues of the matrices $\mathrm{A} + \mathrm{BF}$ and $\mathrm{A} + \mathrm{HC}$. When these values are fixed to be equal to the optimal eigenvalues associated to the solution of the CARE equation and the FARE equation, the factorization is said to be unitary coprime factorization.

Given the system $P(s) = N(s)M^{-1}(s)$, indicated with P_2 the CARE solution

$$\mathrm{A}^T\mathrm{P}_2 + \mathrm{P}_2\mathrm{A} - \mathrm{P}_2\mathrm{B}\mathrm{B}^T\mathrm{P}_2 + \mathrm{C}^T\mathrm{C} = 0$$

and with P_1 the FARE solution

$$\mathrm{A}\mathrm{P}_1 + \mathrm{P}_1\mathrm{A}^T - \mathrm{P}_1\mathrm{C}^T\mathrm{C}\mathrm{P}_1 + \mathrm{B}\mathrm{B}^T = 0$$

in the unitary right coprime factorization we fix as eigenvalues of $M(s)$ and $N(s)$ the optimal eigenvalues. So we choose $\mathrm{F} = -\mathrm{B}^T\mathrm{P}_2$ in the equations (13.5) and (13.6). Concerning $X(s)$ and $Y(s)$, we choose in the equations (13.7) and (13.8) $\mathrm{H} = -\mathrm{P}_1\mathrm{C}^T$. In this way we get:

$$N^T(-s)N(s) + M^T(-s)M(s) = I$$

Unitary factorization is also defined for left factorized MIMO systems $P(s) = \tilde{M}^{-1}(s)\tilde{N}(s)$. In this case we talk about left unitary coprime factorization if

$$\tilde{N}(s)\tilde{N}^T(-s) + \tilde{M}(s)\tilde{M}^T(-s) = I$$

with

$$\tilde{M}(s) = \left[\begin{array}{c|c} A^* & C \\ \hline H^* & I \end{array}\right] \tag{13.17}$$

$$\tilde{N}(s) = \left[\begin{array}{c|c} A^* & C \\ \hline B & 0 \end{array}\right] \tag{13.18}$$

and $A^* = A + H^*C$ and $H^* = -P_1C^T$.

13.6 Parameterization in presence of uncertainty

Consider now the case in which in the process $P(s)$ there is an uncertainty: $P(s) = P_0(s) + r(s)$. $P_0(s)$ is the nominal transfer function, while $r(s)$ is the additive structural uncertainty. So we have $|P(j\omega) - P_0(j\omega)| < |r(j\omega)|$.

Let's define with $T(s) = \frac{C(s)P_0(s)}{1+C(s)P_0(s)}$ the closed-loop transfer function. The techniques analyzed above permit us to parameterize the class of compensators $C(s)$ that make $T(s)$ asymptotically stable. Now we set the objective to find the class of compensators that stabilize the closed-loop system in presence of structural uncertainties.

Often the uncertainty is defined considering the following normalization:

$$\frac{|P(j\omega) - P_0(j\omega)|}{|P_0(j\omega)|} < \frac{|r(j\omega)|}{|P_0(j\omega)|}$$

or in the s domain

$$\frac{P(s) - P_0(s)}{P_0(s)} = m(s)$$

with $m(s) = \frac{r(s)}{P_0(s)}$. So we want to find the class of compensators $C(s)$ that stabilizes any process $P(s) = P_0(s)[1 + m(s)]$. To do this we have to impose also that the zeros of $1 + C(s)P(s)$ (i.e., the closed-loop poles) are zeros with strictly negative real part.

Since

$$1 + C(s)P(s) = 1 + C(s)P_0(s)(1 + m(s)) =$$

$$= (1 + C(s)P_0(s))[1 + m(s)\frac{C(s)P_0(s)}{1 + C(s)P_0(s)}]$$

$1 + C(s)P(s)$ is given by the product of two terms: $1 + C(s)P_0(s)$ and $1 +$

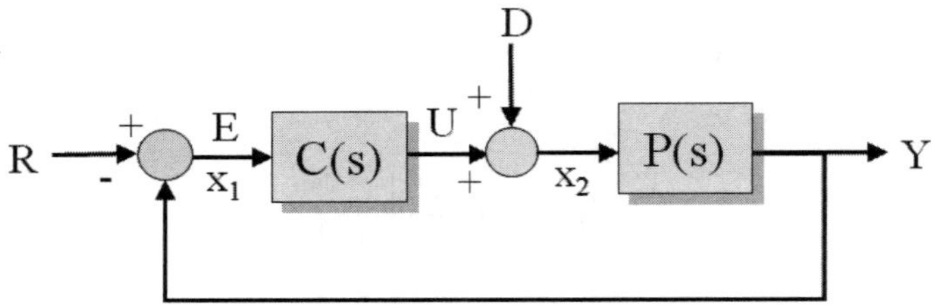

FIGURE 13.4
Feedback control scheme.

$m(s)T(s)$. The first term $1+C(s)P_0(s)$ certainly has roots with negative real part, because it is the denominator of the closed-loop transfer function $T(s)$. The second term $1+m(s)T(s)$ depends on the closed-loop transfer function and on the uncertainty $m(s)$.

If $|m(s)T(s)|<1$ for s with positive real part, we have the guarantee that $1+m(s)T(s)$ never annuls for s with positive real part. Therefore, this term does not have roots that would lead the system to instability.

Consider the scheme in Figure 13.4 and consider the transfer functions:

$$\begin{bmatrix} \frac{X_1(s)}{R(s)} & \frac{X_1(s)}{D(s)} \\ \frac{X_2(s)}{R(s)} & \frac{X_2(s)}{D(s)} \end{bmatrix} = \begin{bmatrix} (1+C(s)P(s))^{-1} & -P(s)(1+C(s)P(s))^{-1} \\ C(s)(1+C(s)P(s))^{-1} & (1+C(s)P(s))^{-1} \end{bmatrix}$$

Consider now the class of compensators defined by $C(s)=\frac{Q(s)}{1-P(s)Q(s)}$. We have seen previously that this class of compensators stabilizes internally the system, when $P(s)$ is stable. The use of this class of compensators for unstable systems requires some stratagems that will be clarified below. For now let us suppose we can use $C(s)=\frac{Q(s)}{1-P(s)Q(s)}$ for which $Q(s)=\frac{C(s)}{1+P(s)C(s)}$. This permits us to express the given transfer functions in function of $P(s)$ and $Q(s)$ (to abbreviate we omit the variable s):

$$\begin{bmatrix} \frac{X_1(s)}{R(s)} & \frac{X_1(s)}{D(s)} \\ \frac{X_2(s)}{R(s)} & \frac{X_2(s)}{D(s)} \end{bmatrix} = \begin{bmatrix} 1-PQ & -P(1-PQ) \\ Q & 1-PQ \end{bmatrix}$$

From this expression we can get the conditions to impose to $Q(s)$ so that the system is internally stable. The term that represents the transfer function from r to x_2 requires that $Q(s)\in\mathbb{H}^\infty$.

Consider now the term $1-PQ$:

$$1-PQ = 1-\frac{N(s)}{D(s)}\frac{Q_1(s)}{Q_2(s)} = \frac{D(s)Q_2(s)-N(s)Q_1(s)}{D(s)Q_2(s)}$$

The fact that $Q(s)\in\mathbb{H}^\infty$ implies that $Q_2(s)$ has roots with negative real part. But $D(s)$ can have roots with positive real part. These roots must be canceled, imposing some interpolation conditions on the numerator of $1-PQ$. From this it follows that, indicated with $\alpha_1, \alpha_2, \ldots, \alpha_n$ the poles of $P(s)$ with positive real part (i.e., the roots of $D(s)$), we must impose that $P(\alpha_i)Q(\alpha_i) = 1$ $\forall \alpha_i$.

Concerning the term $P(1-PQ)$, we will clarify below that the necessary stratagem to treat unstable systems is that also this transfer function is stable.

The last condition to impose derives from the considerations on the process disturbances described above. We must impose that $|m(s)T(s)| < 1$ $\forall s$ with non-negative real part. But since $T(s) = \frac{C(s)P_0(s)}{1+C(s)P_0(s)}$, $Q(s) = \frac{C(s)}{1+P_0(s)C(s)}$ and $m(s) = \frac{r(s)}{P_0(s)}$, we have $m(s)T(s) = r(s)Q(s)$. If $s = j\omega$ then we have to impose that $|r(j\omega)Q(j\omega)| < 1$ $\forall \omega$. The condition is then that $r(s)Q(s)$ is strictly bounded real.

In conclusion, the conditions to impose to $Q(s)$ to obtain the class of controllers $C(s) = \frac{Q(s)}{1-P(s)Q(s)}$ that stabilize internally the process $P(s) = P_0(s) + r(s)$ are:

1. $Q(s) \in H_\infty$;
2. $P(\alpha_i)Q(\alpha_i) = 1$ for any unstable pole α_i of $P(s)$;
3. $|r(j\omega)Q(j\omega)| < 1$ $\forall \omega$.

Note that to apply the condition $P(\alpha_i)Q(\alpha_i) = 1$ since $P(s)$ is unstable requires a particular stratagem. The so-called Blaschke product is used.

Consider for example an unstable system $P(s) = \frac{1}{(s-1)(s-2)}$ and consider also the system $\tilde{P}(s) = \frac{1}{(s+1)(s+2)}$. The system $\tilde{P}(s)$ can be obtained from the product (called a Blaschke product) of the system $P(s)$ and the all-pass system: $B(s) = \frac{(s-1)(s-2)}{(s+1)(s+2)}$.

The idea is to apply the conditions discussed previously to the system $\tilde{P}(s)$ taking into account the unstable poles of $P(s)$. Applying the conditions 1)-3) to the system $\tilde{P}(s)$ we obtain $\tilde{Q}(s)$. So we get $Q(s) = B(s)\tilde{Q}(s)$. In this way $\tilde{Q}(s) \in \mathbb{H}^\infty$ for construction, but also $Q(s) \in \mathbb{H}^\infty$ as a product of two stable functions (in fact, also $B(s) \in \mathbb{H}^\infty$).

Note at this point that $\tilde{P}(s)\tilde{Q}(s) = P(s)Q(s)$. So, in this way it is possible to build a transfer function $Q(s)$ which satisfies the conditions 1)-3) but starting from a $\tilde{P}(s) \in \mathbb{H}^\infty$. The condition $\tilde{P}(\alpha_i)\tilde{Q}(\alpha_i) = 1$ with α_i indicating the poles of $P(s)$ with positive real part in fact can be applied without having the problems discussed previously.

Finally note that, since in this context designing strictly bounded real functions is more complicated than designing positive real functions, often we design $Q(s)$ to obtain a positive real function and then we apply the scattering matrix to obtain a bounded real function.

13.7 Exercises

1. Given the system $G(s) = \frac{s+5}{s^2+2s+2}$ determine the class of stabilizing compensators. Then determine a closed-loop compensator that assures zero error of the unit step response.
2. Given the system $G(s) = \frac{s+5}{s^2-2s+2}$ determine the class of stabilizing compensators. Then determine a closed-loop compensator that assures zero error of the unit step response.
3. Given the system $G(s) = \frac{1}{s-2}$ determine the class of the stable and stabilizing compensators. Then determine a closed-loop compensator that assures a pole in $p = -5$.
4. Given the system $G(s) = \frac{s-1}{(s-2)(s-5)}$ determine the class of stable and stabilizing compensators.
5. Given the system with transfer function $G(s) = \frac{4}{s-3}$ determine the coprime factorization. Then determine the class of stabilizing compensators with unit step response equal to 1. Finally, calculate the energy associated to the impulse response for the closed-loop system.
6. Determine the right and left unitary coprime factorization of the system $G(s) = \frac{1}{(s-1)^2}$.
7. Calculate the unitary coprime factorization of the system with transfer function $G(s) = \frac{s+1}{s^2+3s+1}$.
8. Calculate a compensator that stabilizes simultaneously the two plants with transfer function $G_1(s) = \frac{1}{s-1}$ and $G_2(s) = \frac{1}{s-2}$.
9. Given the systems

$$\begin{cases} \dot{\mathbf{x}} = \mathrm{A}_i\mathbf{x} + \mathrm{B}_i\mathbf{u} \\ \mathbf{y} = \mathrm{C}_i\mathbf{x} \end{cases}$$

with $\mathrm{A}_i = \begin{bmatrix} \alpha_i & 0 & 0 \\ 0 & -1 & 0 \\ 0 & 0 & -2 \end{bmatrix}$; $\mathrm{B}_i = \mathrm{C}_i^T = \begin{bmatrix} 1 \\ -1 \\ \theta_i \end{bmatrix}$ and with $\alpha_i = i$ e $\theta_i = i + 1$ for $i = \{1, 2\}$, determine if it is possible to simultaneously stabilize them and if so, design the linear state regulator and observer so the two systems are asymptotically stable. Verify the result.

10. Calculate, if possible, a control law $u = -Kx$ that can simultaneously stabilize the two systems with state-space matrices:

$$A_1 = \begin{bmatrix} -1 & 0 & 0 & 0 & 0 \\ 0 & 2 & 0 & 0 & 0 \\ 0 & 0 & \sqrt{5} & 0 & 0 \\ 0 & 0 & 0 & -3 & 0 \\ 0 & 0 & 0 & 0 & -4 \end{bmatrix}; A_2 = \begin{bmatrix} 1 & 0 & 0 & 0 & 0 \\ 0 & \sqrt{7} & 0 & 0 & 0 \\ 0 & 0 & -4 & 0 & 0 \\ 0 & 0 & 0 & -5 & 0 \\ 0 & 0 & 0 & 0 & -7 \end{bmatrix};$$

$$B_1^T = B_2^T = \begin{bmatrix} 1 & 1 & 1 & 1 & 0 \end{bmatrix}^T.$$

Recommended essential references

Chapter 1

1. G. Franklin, J. D. Powell, A. Emami-Naeini, *Feedback Control of Dynamic Systems*, 5th Edition, Prentice Hall, Englewood Cliffs, NJ, 2004.
2. J. C. Doyle, B. A. Francis, A. R. Tannenbaum, *Feedback Control Theory*, Macmillan Publishing Company, New York, 1992.
3. W. S. Levine, *The Control Handbook*, CRC Press and IEEE Press, Boca Raton, 1996.
4. S. P. Bhattacharyya, H. Chapellat, L. H. Keel, *Robust Control. The Parametric Approach*, Prentice Hall PTR, 2000.
5. K. Zhou, J. C. Doyle *Essentials of Robust Control*, Prentice Hall PTR, 1999.

Chapter 2

1. T. Kailath, *Linear Systems*, Prentice Hall, Englewood Cliffs, NJ, 1979.
2. D. G. Schultz, J. L. Melsa, *State Functions and Linear Control Systems*, McGraw-Hill, New York, 1967.
3. V. L. Kharitonov, "Asymptotic stability of an equilibrium position of a family of systems of linear differential equations," *Differential'nye Uraveniya*, vol. 14, 1978, pp. 1483–1485.
4. R. C. Dorf, R. H. Bishop, *Modern Control Systems*, 12th Edition, Prentice Hall, Englewood Cliffs, NJ, 2010.
5. W. S. Levine, *The Control Handbook*, CRC Press and IEEE Press, Boca Raton, 1996.

Chapter 3

1. H. Kwakernaak, R. Sivan, *Linear Optimal Control Systems*, John Wiley & Sons, 1972.
2. W. J. Rugh, *Linear System Theory*, Prentice Hall, Englewood Cliffs, NJ, 1966.
3. T. Kailath, *Linear Systems*, Prentice Hall, Englewood Cliffs, NJ, 1979.
4. C. Piccardi, S. Rinaldi, *I sistemi lineari: teoria, modelli, applicazioni*, Città Studi, Biella, 1998.
5. N. Wiener, *Nonlinear Problems in Random Theory*, The Massachusetts Institute of Technology, 1958.

Chapter 4

1. G. H. Golub, C. F. Van Loan, *Matrix Computations*, The Johns Hopkins University Press, Baltimore, MD, 1996.
2. T. Kailath, *Linear Systems*, Prentice Hall, Englewood Cliffs, NJ, 1979.
3. W. S. Levine, *The Control Handbook*, CRC Press and IEEE Press, Boca Raton, 1996.

Chapter 5

1. K. Zhou K., J. C. Doyle, K. Glover, *Robust and Optimal Control*, Prentice Hall, Upper Saddle River, 1996.
2. F. W. Fairman, *Linear Control Theory: The State Space Approach*, John Wiley & Sons, West Sussex, England, 1998.
3. R. A. Roberts, C. T. Mullis, *Digital Signal Processing*, Addison-Wesley Publishing Company, 1987.
4. C. K. Chui, G. Chen, *Discrete H_∞ Optimization*, 2nd Edition, Springer, London, 1997.

Chapter 6

1. B. Moore, "Principal component analysis in linear systems: Controllability, observability, and model reduction," *IEEE Trans. Automatic Control*, vol. 26, no. 1, 1981, pp. 17–32.
2. G. Obinata, B. D. O. Anderson, *Model Reduction for Control System Design*, Springer-Verlag, London, 2001.
3. M. Green, D. J. N. Limebeer, *Linear Robust Control*, Prentice Hall, Englewood Cliffs, NJ, 1995.

Chapter 7

1. F. Fagnani, J. Willems, "Representations of symmetric linear dynamical systems," *SIAM Journal on Control and Optimization archive*, vol. 31, no. 5, 1993.
2. L. Fortuna, A. Gallo, C. Guglielmino, G. Nunnari, "On the solution of a nonlinear matrix equation for MIMO realizations," *Systems and Control Letters*, vol. 11, 1988, pp. 79–82.
3. L. Fortuna, G. Nunnari, A. Gallo, *Model Order Reduction Techniques With Applications in Electrical Engineering*, Springer-Verlag, London, 1992.

Chapter 8

1. B. D. O. Anderson, J.B. Moore, *Optimal Control. Linear Quadratic Methods*, Prentice-Hall, Englewoods Cliff, NJ, 1989.
2. H. Kwakernaak, R. Sivan, *Linear Optimal Control Systems*, John Wiley & Sons, 1972.
3. T. Kailath, *Linear Systems*, Prentice Hall, Englewood Cliffs, NJ, 1979.
4. G. C. Goodwin, S. F. Graebe, M. E. Salgado, *Control Systems Design*, Valparaiso, 2000.
5. A. Sinha, *Linear Systems: Optimal and Robust Control*, CRC Press, Boca Raton, 2007.

6. D. Kleinman, "On an iterative technique for Riccati equation computations," *IEEE Trans. Automatic Control*, vol. 13, no. 1, 1968, pp. 114–115.
7. K. J. Astrom, R. M. Murray, *Feedback Systems. An Introduction for Scientists and Engineers*, Princeton University Press, Princeton, NJ, 2008.
8. W. S. Levine, *The Control Handbook*, CRC Press and IEEE Press, Boca Raton, 1996.

Chapter 9

1. E. A. Jonckheere, L. M. Silverman, "A new set of invariant for linear systems. Applications to reduced order compensator design," *IEEE Trans. Automatic Control*, vol. 26, no. 1, 1981, pp. 17–32.
2. L. Fortuna, G. Nunnari, A. Gallo, *Model Order Reduction Techniques With Applications in Electrical Engineering*, Springer-Verlag, London, 1992.
3. L. Fortuna, G. Muscato, G. Nunnari, "Closed-loop balanced realizations of LTI discrete-time systems," *Journal of the Franklin Institute*, vol. 330, no. 4, 1993, pp. 695-705.
4. K. J. Astrom, R. M. Murray, *Feedback Systems. An Introduction for Scientists and Engineers*, Princeton University Press, Princeton, NJ, 2008.

Chapter 10

1. B. D. O. Anderson, S. Vongpanitlerd, *Network Analysis and Synthesis: A Modern Systems Theory Approach*, Dover Publications Inc., New York, 2006.
2. S. Boyd, L. El Ghaoui, E. Feron, V. Balakrishnan, *Linear Matrix Inequalities in System and Control Theory*, SIAM Studies in Applied Mathematics, Philadelphia, PA, 1994.

Chapter 11

1. R. T. Stefani, B. Shahian, C. J. Savant, G. H. Hostetter, *Design of Feedback Control Systems*, Oxford Series in Electrical and Computer Engineering, Oxford University Press, Oxford, 2001.
2. P. Dorato, R. K. Yedavalli, *Recent Advances in Robust Control*, IEEE Press, New York, 1990.
3. S. Boyd, V. Balakrishnan, P. Kabamba, "A Bisection Method for Computing the H_∞ Norm of a Transfer Matrix and Related Problems," *Math. Control Signals Systems*, vol. 2, 1989, pp. 207–219.
4. P. Dorato, L. Fortuna, G. Muscato, *Robust Control for Unstructured Perturbations-An Introduction*, Lecture Notes in Control and Information Sciences, Springer-Verlag, London, 1992.
5. R. Chiang, M. G. Safonov, G. Balas, A. Packard, *Robust Control Toolbox*, 3rd Edition, The Mathworks Inc., Natick, MA, 2007.
6. B. A. Francis, *A course in H_∞ control theory*, Lecture Notes in Control and Information Sciences, Springer-Verlag, Berlin, 1987.
7. J. C. Doyle, K. Glover, P. P. Khargonekar, B. A. Francis, "State-space solutions to standard H_2 and H_∞ control problems", *IEEE Trans. Automatic Control*, vol. 34, 1989, pp. 831–847.
8. W. S. Levine, *The Control Handbook*, CRC Press and IEEE Press, Boca Raton, 1996.

Chapter 12

1. S. Boyd, L. El Ghaoui, E. Feron, V. Balakrishnan, *Linear Matrix Inequalities in System and Control Theory*, SIAM Studies in Applied Mathematics, Philadelphia, PA, 1994.
2. R. E. Skelton, T. Iwasaki, K. M. Grigoriadis, *A Unified Algebraic Approach to Linear Control Design*, Taylor & Francis, 1998.
3. P. Gahinet, A. Nemirovski, A. J. Laub, M. Chilali, *LMI Control Toolbox. For Use with MATLAB®*, The Mathworks Inc., Natick, MA, 1995.
4. Y. Nesterov and A. Nemirovski, *Interior Point Polynomial Algorithms in Convex Programming*, Studies in Applied Mathematics, SIAM, 1993.

5. U. Helmke, J. B. Moore, *Optimization and Dynamical Systems*, Springer-Verlag, London, 1996.

6. L. Fortuna, M. Frasca, M. G. Xibilia, *Chua's Circuit Implementations: Yesterday, Today and Tomorrow*, World Scientific, Singapore, 2009.

7. W. S. Levine, *The Control Handbook*, CRC Press and IEEE Press, Boca Raton, 1996.

8. L. Fortuna, A. Gallo, G. Nunnari, "A self-tuning adaptive control implementation by using a transputer-based parallel architecture", *Trans. Inst. MC*, vol. 12, 1990, pp. 156–164.

Chapter 13

1. D. C. Youla, J. J. Bongiorno, C. N. Lu, *Single-loop Feedback Stabilization of Linear Multivariable Dynamical Plants*, American Math Society, Colloquium Publications, vol. XX, Providence, RI, 1956.

2. D. C. Youla, H. A. Jabr, J. J. Bongiorno, "Modern Wiener-Hopf design of optimal controllers. Part I: the single-input-output case," *IEEE Trans. Automatic Control*, vol. 10, 1974, pp. 159–173.

3. P. Dorato, L. Fortuna, G. Muscato, *Robust Control for Unstructured Perturbations-An Introduction*, Lecture Notes in Control and Information Sciences, Springer-Verlag, London, 1992.

4. J. C. Doyle, B. A. Francis, A. R. Tannenbaum, *Feedback Control Theory*, Macmillan Publishing Company, New York, 1992.

5. B. D. O. Anderson, J.B. Moore, *Optimal Control. Linear Quadratic Methods*, Prentice-Hall, Englewood Cliffs, NJ, 1989.

6. W. S. Levine, *The Control Handbook*, CRC Press and IEEE Press, Boca Raton, 1996.

Appendix A. Norms

In this Appendix the main norms used in the text or, in any case, significant for the control theory, are summarized.

Norm of a vector

The norm of a vector $\mathbf{x} \in \mathbb{C}^n$ is a function $f : \mathbb{C}^n \to \mathbb{R}$ satisfying the following properties:

- $f(\mathbf{x}) \geq 0$;
- $f(\mathbf{x}) = 0 \leftrightarrow \mathbf{x} = 0$;
- $f(\mathbf{x} + \mathbf{y}) \leq f(\mathbf{x}) + f(\mathbf{y})$ con $\mathbf{x}$ e $\mathbf{y} \in \mathbb{C}^n$;
- $f(\alpha \mathbf{x}) = |\alpha| f(\mathbf{x})$ con $\alpha \in \mathbb{R}$.

Let $\mathbf{x} = \begin{bmatrix} x_1 \\ x_2 \\ \vdots \\ x_n \end{bmatrix}$, the 1-norm is:

$$\|\mathbf{x}\|_1 = \sum_{i=1}^{n} |x_i| \tag{A.1}$$

The 2-norm is instead defined by:

$$\|\mathbf{x}\|_2 = \left(\sum_{i=1}^{n} |x_i|^2 \right)^{\frac{1}{2}} \tag{A.2}$$

The 1-norm and the 2-norm are special cases of a more general family of norms, the p-norm of a vector:

$$\|\mathbf{x}\|_p = \left(\sum_{i=1}^{n} |x_i|^p \right)^{\frac{1}{p}} \tag{A.3}$$

Another commonly used norm of the p-norm family is the ∞-norm defined as:

$$\|\mathbf{x}\|_\infty = \max_i |x_i| \tag{A.4}$$

Norm of a matrix

By considering the role of matrices as linear operators, the norm of a matrix can be defined extending the norm of a vector to matrices. The norm is said to be an *induced norm*, since it depends on the choice of the vector norm. The p-norm (induced by the vector p-norm) is defined as follows:

$$\|\mathrm{A}\|_p = \sup_{\mathbf{x}\neq 0} \frac{\|\mathrm{A}\mathbf{x}\|_p}{\|\mathbf{x}\|_p} = \max_{\|\mathbf{x}\|=1} \|\mathrm{A}\mathbf{x}\|_p \tag{A.5}$$

The most commonly used norms are the 1-norm, the 2-norm and the ∞-norm, defined as follows:

$$\|\mathrm{A}\|_1 = \max_j \sum_{i=1}^{n} |a_{ij}| \tag{A.6}$$

$$\|\mathrm{A}\|_2 = \sigma_1(\mathrm{A}) = \max_i \lambda_i(\mathrm{A}^*\mathrm{A}) \tag{A.7}$$

$$\|\mathrm{A}\|_\infty = \max_i \sum_{j=1}^{n} |a_{ij}| \tag{A.8}$$

Additional norms (which are not induced norms) can be defined. We report only the Frobenius norm, which is defined as follows:

$$\|\mathrm{A}\|_F = \left(\sum_{j=1}^{n} \sum_{i=1}^{n} |a_{ij}|^2 \right)^{\frac{1}{2}} \tag{A.9}$$

Norm of a scalar signal

Given a signal $g(t) \in \mathbb{R}$ with $t \in \mathbb{R}$, the 1-norm, the 2-norm and the ∞-norm are defined as follows:

$$\|g(t)\|_1 = \int_{-\infty}^{+\infty} |g(t)| dt \tag{A.10}$$

$$\|g(t)\|_2 = \left(\int_{-\infty}^{+\infty} |g(t)|^2 dt \right)^{\frac{1}{2}} \tag{A.11}$$

$$\|g(t)\|_\infty = \sup_{t \in \mathbb{R}} |g(t)| \tag{A.12}$$

Norm of a vector signal

For a vector-valued signal $\mathbf{g}(t) = \begin{bmatrix} g_1(t) \\ g_2(t) \\ \vdots \\ g_n(t) \end{bmatrix}$, the 1-norm, the 2-norm and the ∞-norm are given by:

$$\|\mathbf{g}(t)\|_1 = \int_{-\infty}^{+\infty} \|\mathbf{g}(t)\|_2 dt \tag{A.13}$$

$$\|\mathbf{g}(t)\|_2 = \left(\int_{-\infty}^{+\infty} \|\mathbf{g}(t)\|_2^2 dt \right)^{\frac{1}{2}} \tag{A.14}$$

$$\|\mathbf{g}(t)\|_\infty = \sup_{t \in \mathbb{R}} \|\mathbf{g}(t)\|_2 \tag{A.15}$$

Norm of a transfer function matrix

For transfer function matrices, usually only two norms are commonly used: the 2-norm and the ∞-norm. Given a transfer function matrix $\mathrm{G}(s)$, and indicated as $\mathrm{g}(t)$ the impulse response, i.e., $\mathrm{g}(t) = \mathcal{L}^{-1}(G(s))$, the 2-norm and the ∞-norm can be defined either in the frequency domain or in the time domain. In the frequency domain the norms are defined as:

$$\|\mathrm{G}(s)\|_2 = \left(\frac{1}{2\pi} \int_{-\infty}^{+\infty} \|\mathrm{G}(j\omega)\|_F^2 d\omega \right)^{\frac{1}{2}} \tag{A.16}$$

$$\|\mathrm{G}(s)\|_\infty = \sup_{\omega \in \mathbb{R}} \sigma_1(\mathrm{G}(j\omega)) \tag{A.17}$$

while in the time domain the two norms are defined by:

$$\|\mathrm{G}(s)\|_2 = \left(\frac{1}{2\pi}\int_{-\infty}^{+\infty} \|\mathrm{g}(t)\|_F^2 dt\right)^{\frac{1}{2}} \tag{A.18}$$

$$\|\mathrm{G}(s)\|_\infty = \sup_{u\neq 0} \frac{\|(\mathrm{G}u)(t)\|_2}{\|u(t)\|_2} \tag{A.19}$$

Appendix B. Algebraic Riccati Equations

Algebraic Riccati equations appear in many problems discussed in this text. In this section we summarize the main algebraic Riccati equations for continuous-time systems.

Algebraic Riccati Equations for optimal control and optimal observer

$$\mathrm{PA} + \mathrm{A}^T\mathrm{P} - \mathrm{PBR}^{-1}\mathrm{B}^T\mathrm{P} + \mathrm{Q} = 0 \tag{B.1}$$

$$\mathrm{A\Pi} + \mathrm{\Pi A}^T - \mathrm{\Pi C}^T\mathrm{M}_v^{-1}\mathrm{C\Pi} + \mathrm{M}_d = 0 \tag{B.2}$$

CARE and FARE for closed loop balancing

$$\mathrm{A}^T\mathrm{P} + \mathrm{PA} - \mathrm{PBB}^T\mathrm{P} + \mathrm{C}^T\mathrm{C} = 0 \tag{B.3}$$

$$\mathrm{A\Pi} + \mathrm{\Pi A}^T - \mathrm{\Pi C}^T\mathrm{C\Pi} + \mathrm{BB}^T = 0 \tag{B.4}$$

CARE and FARE for closed loop balancing for not-strictly proper systems

$$\mathrm{A}^T\mathrm{P} + \mathrm{PA} - (\mathrm{PB} + \mathrm{C}^T\mathrm{D})(\mathrm{I} + \mathrm{D}^T\mathrm{D})^{-1}(\mathrm{B}^T\mathrm{P} + \mathrm{D}^T\mathrm{C}) + \mathrm{C}^T\mathrm{C} = 0 \tag{B.5}$$

$$A\Pi + \Pi A^T - (\Pi C^T + BD^T)(I + DD^T)^{-1}(C\Pi + DB^T) + BB^T = 0 \tag{B.6}$$

Algebraic Riccati Equations for positive real systems

$$PA + A^T P = -(-PB + C^T)(D + D^T)^{-1}(-PB + C^T)^T \tag{B.7}$$

$$\Pi A^T + A\Pi = -(-\Pi C^T + B)(D + D^T)^{-1}(-\Pi C^T + B)^T \tag{B.8}$$

Algebraic Riccati Equations for bounded real systems

$$\begin{aligned} &P(A + B(I - D^T D)^{-1} D^T C) + (A^T + C^T D(I - D^T D)^{-1} B^T)P + \\ &+ PB(I - D^T D)^{-1} B^T P + C^T D(I - D^T D)^{-1} D^T C + C^T C = 0 \end{aligned} \tag{B.9}$$

$$\begin{aligned} &P(A^T + C^T(I - DD^T)^{-1} DB^T) + (A + BD^T(I - DD^T)^{-1} C)P + \\ &+ PC^T(I - DD^T)^{-1} CP + BD^T(I - DD^T)^{-1} DB^T + BB^T = 0 \end{aligned} \tag{B.10}$$

Algebraic Riccati Equations for strictly proper bounded real systems

$$PA + A^T P + PBB^T P + C^T C = 0 \tag{B.11}$$

$$PA^T + AP + PC^T CP + BB^T = 0 \tag{B.12}$$

Algebraic Riccati Equations for H_∞ control

$$\begin{aligned}&(\mathrm{A}-\mathrm{B}_2\tilde{\mathrm{D}}_{12}\mathrm{D}_{12}^T\mathrm{C}_1)\mathrm{X}_\infty+\mathrm{X}_\infty(\mathrm{A}-\mathrm{B}_2\tilde{\mathrm{D}}_{12}\mathrm{D}_{12}^T\mathrm{C}_1)^T+\\&+\mathrm{X}_\infty(\gamma^{-2}\mathrm{B}_1\mathrm{B}_1^T-\mathrm{B}_2\tilde{\mathrm{D}}_{12}\mathrm{B}_2^T)\mathrm{X}_\infty+\tilde{\mathrm{C}}_1^T\tilde{\mathrm{C}}_1=0\end{aligned} \tag{B.13}$$

$$\begin{aligned}&(\mathrm{A}-\mathrm{B}_2\tilde{\mathrm{D}}_{12}\mathrm{D}_{12}^T\mathrm{C}_1)^T\mathrm{X}_\infty+\mathrm{X}_\infty(\mathrm{A}-\mathrm{B}_2\tilde{\mathrm{D}}_{12}\mathrm{D}_{12}^T\mathrm{C}_1)+\\&+\mathrm{X}_\infty(\gamma^{-2}\mathrm{C}_1^T\mathrm{C}_1-\mathrm{C}_2^T\tilde{\mathrm{D}}_{21}\mathrm{C}_2)\mathrm{X}_\infty+\tilde{\mathrm{B}}_1\tilde{\mathrm{B}}_1^T=0\end{aligned} \tag{B.14}$$

with $\tilde{\mathrm{C}}_1=(\mathrm{I}-\mathrm{D}_{12}\tilde{\mathrm{D}}_{12}\mathrm{D}_{12}^T)\mathrm{C}_1$, $\tilde{\mathrm{B}}_1=\mathrm{B}_1(\mathrm{I}-\mathrm{D}_{21}^T\tilde{\mathrm{D}}_{21}\mathrm{D}_{21})$, $\tilde{\mathrm{D}}_{12}=(\mathrm{D}_{12}^T\mathrm{D}_{12})^{-1}$ and $\tilde{\mathrm{D}}_{21}=(\mathrm{D}_{21}\mathrm{D}_{21}^T)^{-1}$.

Cross Riccati equations for optimal control and optimal observer (for symmetrical systems)

For symmetrical systems, cross-Riccati equations can be defined. As an example, we report those related to optimal control and optimal observer:

$$\mathrm{AP}^*+\mathrm{P}^*\mathrm{A}-\mathrm{P}^*\mathrm{BCP}^*+\mathrm{BC}=0 \tag{B.15}$$

Index